Mind and Variability

Recent Titles in
Human Evolution, Behavior, and Intelligence

A People That Shall Dwell Alone: Judaism as a Group Evolutionary Strategy
Kevin MacDonald

Dysgenics: Genetic Deterioration in Modern Populations
Richard Lynn

Darwinism, Dominance, and Democracy: The Biological Bases of Authoritarianism
Albert Somit and Steven A. Peterson

Why Race Matters: Race Differences and What They Mean
Michael Levin

The Evolution of Love
Ada Lampert

The *g* Factor
Arthur R. Jensen

Sex Linkage of Intelligence: The X-Factor
Robert Lehrke

Separation and Its Discontents: Toward an Evolutionary Theory of Anti-Semitism
Kevin MacDonald

The Biological Origins of Art
Nancy E. Aiken

The Culture of Critique: An Evolutionary Analysis of Jewish Involvement in Twentieth-Century Intellectual and Political Movements
Kevin MacDonald

Relating in Psychotherapy: The Application of a New Theory
John Birtchnell

The Evolution of the Psyche
D. H. Rosen and M. C. Luebbert, editors

Mind and Variability

Mental Darwinism, Memory, and Self

Patrick McNamara

Human Evolution, Behavior, and Intelligence
Seymour W. Itzkoff, *Series Editor*

PRAEGER

Westport, Connecticut
London

Library of Congress Cataloging-in-Publication Data

McNamara, Patrick, 1956–
Mind and variability : mental Darwinism, memory, and self / Patrick McNamara.
p. cm.—(Human evolution, behavior, and intelligence, ISSN 1063–2158)
Includes bibliographical references and index.
ISBN 0–275–96383–7 (alk. paper)
1. Memory. 2. Brain—Evolution. 3. Identity (Psychology) 4. Genetic psychology. I. Title. II. Series.
BF371.M385 1999
153.1′2—dc21 98–31365

British Library Cataloguing in Publication Data is available.

Library of Congress Catalog Card Number: 98–31365
ISBN: 0–275–96383–7
ISSN: 1063–2158

First published in 1999

Praeger Publishers, 88 Post Road West, Westport, CT 06881
An imprint of Greenwood Publishing Group, Inc.
www.praeger.com

Printed in the United States of America

The paper used in this book complies with the Permanent Paper Standard issued by the National Information Standards Organization (Z39.48–1984).

P

Copyright Acknowledgments

The author and publisher gratefully acknowledge permission for use of the following materials:

Patrick McNamara. "Bergson's Theory of Dreaming." *Dreaming: Journal of the Association for the Study of Dreams* 6, no. 3 (1996): 173–186. Copyright by the Association for the Study of Dreams. Originally published by Human Sciences Press, New York.

Patrick McNamara. "Bergson's 'Matter and Memory' and Modern Selectionist Theories of Memory." *Brain and Cognition* 30 (1996): 215–231. Reprinted by permission of Academic Press, Orlando, Florida.

In order to keep this title in print and available to the academic community, this edition was produced using digital reprint technology in a relatively short print run. This would not have been attainable using traditional methods. Although the cover has been changed from its original appearance, the text remains the same and all materials and methods used still conform to the highest book-making standards.

In memory of MM & MTM

Table of Contents

Preface

Of all the faculties of the Mind, memory seems so obviously to require an "instructionist" account that it seems reckless to question it; yet that is what I do in this book. On the instructionist account, memory occurs when information is transferred from the environment into the "Mind." But suppose (with the Darwinian psychologists) that the Mind is already teeming with information of all kinds, including those images we call memories, and that all that the environment does is to promote a select few of these images? What then? That is the question I seek to answer in this book. Readers can judge for themselves whether importing Darwinian principles into the study of the Mind has been worthwhile. I, of course, believe it has. The Darwinian approach is interesting because it requires us to enlarge our conceptions of Mind while simultaneously showing us that its fundamental operating principles boil down to three simple processes (see Chapter 1). Not every scientist is ready or willing to entertain these Darwinian temptations when it comes to studying the Mind, so I have appreciated the work of those investigators (many of whom are cited in this text) who have published their findings, speculations, and ideas before me. I also wish to thank all of the following friends and colleagues for their help and inspiration along the way: Carol Beasley; Jill Clark; Ray Durso; Ernest Hartmann; Sarah Heller; David Kemmerer; Reema Khan; Susan Lanzoni; Misia Landau; Cynthia Merillat; David Price; Jim Whitlock; Michael Zborowski; the Putnam Road gang; the three Chapwomen: Carol, Cathy and Ginger; Lisette Coly and Joanne McMahon of the Eileen J. Garrett

Library of the Parapsychology Foundation in New York; and my teachers in Boston: Martin Albert, Laird Cermak, Loraine Obler, Marlene Oscar-Berman, and Edgar Zurif. Naturally, none of these should be held responsible for the errors or interpretations found in this book.

Introduction

We moderns and post-moderns are obsessed with the related themes of memory and identity, and rightfully so. We intuitively realize that much is at stake when it comes to memory and identity. In no two other areas are the personal and public realms of experience so densely intertwined. To say as much is to claim that the stuff of our experience, our pleasures and pains, our sufferings and triumphs, our breakthroughs and our failures are all bound up with the things we call memory and identity. What is the nature of these two protean phenomena? How are they interrelated? How do they give rise to our experiences?

The eighteenth-century Italian philosopher, Giambattista Vico, suggested that memory was composed of three processes: memoria, fantasia and ingegno. Memoria refers to the everyday act of remembering "things"; fantasia refers to the "imagination when it alters or imitates them . . ."; ingegno refers to "invention when it gives them a new turn or puts them into proper arrangement and relationship" (*The New Science*, 1744). Vico, in other words, places imaginative and inventive activity at the heart of recollective experience. This book represents an extended argument in favor of Vico's conception of memory. If Vico is right that imaginative activity and memory are indissolubly linked, then what we call personal identity might also be an imaginative creation. It might seem strange that I rely on Darwin to elucidate Vico's arguments on memory. Yet Darwinian principles of selection require a pool of variable elements from which to select and this pool of variability in the realm of the Mind is the imagination. As

William James argued, "Mental Darwinism" places the imagination at the center of human experience and thus Vico is vindicated by Darwin.

In the following pages I present a "selectionist" approach to memory, as I believe it captures Vico's fundamental insights about the nature of memory. When Charles Darwin showed that the vast number of organisms that have existed over equally vast epochs could be fruitfully explained by invoking a very simple set of selectionist principles, he initiated a revolution in the sciences and in thought that continues to this day. Darwin showed that given (a) a generator of variability around some particular trait or ability, (b) a selective agent that acts to preserve a subset of that initial variability, (c) amplification of the selected traits into the population, and (d) further generation of variants from that preserved or selected set of traits, you will get evolutionary development and an increasing "fit" of the trait or ability to a given ecological niche.

Some twenty to thirty years after Darwin published his theory of natural selection, William James in America and Henri Bergson in France attempted to apply Darwinian principles to problems of memory and identity. Although these two philosophers were celebrated in their time and are still respected names in psychological research today, their contributions to memory theory have gone unappreciated. I have tried to give them their due in this book by adhering closely to their respective views on selectionist memory theory. Although speculative in tone, I try to ground my interpretations and developments of James's and Bergson's views in modern neuropsychological and experimental findings.

In Chapter 1, I introduce the fundamentals of selectionist theory in general, and why I believe it is a particularly promising theory for problems of memory and identity. In Chapter 2, I summarize the shortcomings of the standard "instructivist" approach to memory. The instructivist approach cannot adequately account for (1) the historical feel or character of memory, (2) the problem of the veridicality of memory, (3) memory's constructive and reconstructive abilities, and (4) the relation between memory and identity. In Chapter 3, I turn to a summary of Henri Bergson's views on memory. This provides the reader with a concrete grasp of what a selectionist theory of remembering might look like. In Chapters 4 and 5, I show that selectionist accounts of cognition rely heavily on inhibitory processes that suppress irrelevant variants of a target memory or image. I provide a summary of the experimental and neuropsychological evidence for such inhibitory and selective processes in cognition. The role of the frontal lobes in episodic remembering is highlighted. In general, an anterior neuropsychological system is designed to select out images from the intrapsychic contributions to the stream of thought, while a posterior system is oriented toward the external sensory portions of the stream. In each act of attention, we initially orient to novelty and generate a new set of images. If we so choose, we bias our selection and attentional processing mechanisms to favor redundant input, which then is more easily transferred to memory. In any case, attentional systems may be set or biased to favor one portion of the stream over another. These biases set up long-term personality dispositions and traits. In Chapters 6 and 7, I summarize William James's ideas on selectionist memory and personal identity. James's central notions of the "stream of thought," and the "hierarchy of selves" are discussed within the context of the problem of variability of mental contents. Variability gives rise to the need for selective attention and selective retention. Excess mental variability imposes two basic attentional strategies on us: We can be well-disposed toward that variability or closed to it. Chapter 8 lays out a theory of the ways in which we relate to the excess mental variability James called the stream of consciousness or thought. Chapter 9 provides a phenomenological account of recollection and touches on the world's mythological imagery of the memory faculty. Chapter 10 provides a selectionist account of dreaming and argues that dreams help to construct the self through a process of editing in autobiographical memory. Chapter 11 presents a selectionist account of the interrelationships among memory, self, and culture and closes with a discussion of Bergson's ideas on the evolution of intelligence in open versus closed societies.

CHAPTER 1

Selection and Memory

MENTAL DARWINISM

"Mental Darwinism" refers to a new approach to the study of the brain and of mental phenomena. As its name implies this new approach applies classical Darwinian principles (as first laid out in *The Origin of Species*, 1859) to problems of Mind and behavior (see, for example, Barkow, Cosmides, Tooby, 1992; Calvin, 1996; Cziko, 1995; Edelman, 1987; Gazzaniga, 1992). One strand of this Darwinian program is the appeal to "selectionist" mechanisms when accounting for brain and mental development of various kinds. In my view, the application of selectionist principles to the human sciences could represent one of the most important developments in psychology and related fields since these fields of study adopted experimental methods in the middle of the nineteenth century. If mental phenomena can be understood within a selectionist framework, we will have a single account of the vast array of "stuff" we call life. While some have abused the reductionist aims of the scientific enterprise by using reduction to ignore the richness and complexity of the phenomena before them, reduction in itself does not require a loss of nuance, complexity or wonder in the account we offer of natural phenomena. By assimilating mental life to the rest of the life sciences, we bring more fundamental laws into the discussion of mental life and the study of mental life is thereby enriched. Adopting the selectionist perspective, furthermore, yields concrete advances in our understanding of psychological functioning.

In this book I develop a selectionist approach to one domain of mental life: that of autobiographical memory. While other investigators in the cognitive sciences have explicitly and successfully used selectionist-inspired approaches to illuminate problems in various areas (e.g., Calvin, 1996; Campbell, 1960, 1990; Cziko, 1995; Mehler and Fox 1985; Piatelli-Palmarini, 1989; Pinker, 1994; Popper, 1963), cognitive memory researchers have not.

Two philosophers working at the turn of the twentieth century thought the selectionist approach appropriate for understanding memory and offered justifications for adopting such an approach to memory and cognition. These men were William James in the United States and Henri Bergson in France. I will lay out James's and Bergson's ideas concerning selection and memory in later chapters. Here I wish only to provide a handful of reasons that suggest that a selectionist approach to memory might indeed be useful. I will need first to briefly identify the essentials of a "selectionist approach" as it is understood by its modern advocates in the brain and cognitive sciences.

Selectionist models of any life process require three distinct processes: a generator of variation that produces an array of elements or traits that become the raw material for the selection process, a selection process (Changeux & Dehaene, 1989) that acts to reduce the number of elements in the initial array (Edelman, 1987; Plotkin, 1994; Ward, 1989), and an amplification process that spreads effects of selection (through a copying mechanism) through the system (Calvin, 1996; Young, 1979). In general, the elements selected are those that happen by chance to be useful in some way for the organism undergoing selection. In addition to these basic requirements, the elements (e.g., traits) subject to selection must exhibit some degree of consistency through time. If the elements changed randomly and continuously, it would not be possible for the "selector" to act selectively with respect to a given trait (Ward, 1989); there would be no selective retention process. This consistency-through-time condition has been called "cumulative retention," or "the principle of accumulation of design." The elements or traits that survive the reduction process are consistently related to the elements that survived previous reduction or selection cycles. These related traits constitute a coherent type since they are in some way related to one another (their relation played a causal role in their selection). These related "types" thereby increase their number (are amplified or copied) in the overall population, and so the population evolves.

Selectionist theories are often contrasted with "instructionist" theories in the behavioral and life sciences. Sporns (1994), for example, put it this way:

> *Instructionist theories in biology stress the transfer of information from one domain (e.g., the environment) to another (e.g. a biological system). The biological system will—by means that differ from case to case-embody or store the information it has received. . . . Selectionist theories on the other hand, generally do not require such information transfer in their construction. Rather they work on the basis of three principles: 1) preexisting variation among components within the biological system, 2) encounter of the biological system with an environment and polling of the system's components, and 3) differential amplification of those components of the system that meet a threshold criterion after encounter.* (p. 5)

Although instructionist accounts of learning or intelligent behavior appear to be quite plausible, selectionist accounts offer alternative and perhaps more parsimonious explanations of the same phenomena. Suppose, for example, that there is a population of monkeys with a range of behavioral traits. Some of these monkeys just by chance have a trait that makes them avoid large animals while other monkeys do not have such a trait. The monkeys with the avoidance trait will avoid lions and leopards simply because they have the trait—not because they remember any previous encounters with these animals. There may never have been any such encounters. The monkeys with the avoidance trait survive to reproduce. They pass the avoidance trait down to their offspring. Meanwhile monkeys who do not possess the avoidance trait are rapidly eaten up by the lions and leopards. The "non-avoidance of large animals" trait is lost to the population (or at least becomes dormant), since none of the monkeys who possessed it survived to reproduce. In the end, what we are left with is a population of monkeys who appear to selectively avoid lions and leopards. We, as new observers on the scene, assume that the monkeys avoid the lions and leopards because they have learned that these animals are dangerous. But, in fact, learning was never really a factor in the development of the monkey population's aversion to lions and leopards. Rather, natural selection was at work. What had appeared to be the workings of learning, or the transfer of information from the environment to the organism, on closer inspection turns out to be the processes of natural selection. The avoidance trait for monkeys was "fit" because the monkeys who possessed it were not preyed on by large predators and so they survived to reproduce.

The monkeys that survived the predation of lions did so not because they had learned that lions were dangerous but because they happened to have possessed the appropriate avoidance-of-lions trait. The pre-existing variability of traits in the monkey population meant that some possessed the appropriate trait (and were therefore fit) and others did not (and therefore perished). This issue of pre-existing variation in populations on which selection "works" is one of the most important distinctions between instructionist theories and selectionist theories—and it is particularly important for memory theory. Consider the fact that virtually all modern theories of memory are instructionist. Memory is conceived as a process of transfer of information from the environment into the organism. Variation is simply considered noise in the instructionist framework. In the selectionist framework (at least in James's and Bergson's version of the selectionist framework), variation (in type and content of mental representations or images) is the pool or the reservoir from which memories are constructed or selected.

Another distinction between instructionist and selectionist systems is that selectionist systems work by selecting information to be remembered from an array of "competitors" or "distractors." Selectionist systems must therefore rely on some form of inhibitory control in order to reduce interference from the distractor items in support of the selection process. Instructionist systems, by contrast, transfer information into the organism without any selection mechanism. All that is required is that the information to be remembered be formatted into a form suitable for long-term memory. No inhibitory mechanism is necessary. Thus, evidence that memory actually utilizes inhibitory mechanisms as part of its normal processing procedures would be consistent with the hypothesis that memory is a selectionist system.

The great American psychologist and philosopher William James was one of the first to grasp the significance of the instructionist versus selectionist approaches for problems of Mind. In one of his many polemics against the misguided "evolutionism" of Herbert Spencer, he stated:

> *Pre-Darwinian philosophers had also tried to establish the doctrine of descent with modification but they all committed the blunder. . . . What preserves an animal with his peculiarity, if it be a useful one, they saw to*

> *be the nature of the environment to which the peculiarity was adjusted. The giraffe with his peculiar neck is preserved by the fact that there are in his environment tall trees whose leaves he can digest. . . . But these philosophers went further and said that the presence of the trees . . . also produced him. They made his neck long by the constant striving they aroused in him to reach up to them. The environment, in short, was supposed, by these writers, to mould the animal by a kind of direct pressure, very much as a seal presses the wax into harmony with itself.* (1956: 222)

The seal and wax analogy is an apt one for the instructionist position and has been used down through the centuries to describe how memory and learning occur. The seal-impression-in-wax analog implies that structure is transferred from an environment into a passive organism. James, on the other hand, argues that one of Darwin's greatest achievements was to refute this instructionist theory of evolutionary change. The environment's role was not to impress structure on a passive organism; rather, it was to preserve or destroy or select pre-existing structures (or adaptations or "peculiarities") with which the organism comes equipped. Pre-existing structures are traits which the animal possesses by chance or as a result of previous selective pressures. They are accidental variations in abilities or capacities that may or may not match current environmental demands. If they do match those demands, they are preserved; if they do not, they are allowed to die off or are destroyed along with the organism that carries them. "Darwin's first achievement was to show the . . . immensely greater mass of changes being produced by internal molecular accidents. . . . His next achievement was to define the true problem with which we have to deal when we study the effects of the visible environment on the animal. That problem is simply this: Is the environment more likely to preserve or to destroy him, on account of this or that peculiarity with which he may be born?" (1956: 223).

With respect to the prospect of applying Darwinian principles to problems of cognition and mind, James had this to say:

> *Here if anywhere, it would seem at first sight as if that school must be right which makes the mind passively plastic, and the environment actively productive of the form and order of its conceptions . . . as if the distinction we have found so useful between the spontaneous variation as the producer of changed forms, and the environment, as their preserver and destroyer, did*

> *not hold in the case of mental progress . . . but . . . as a matter of fact the new conceptions, emotions, and active tendencies which evolve are originally produced in the shape of random images, fancies, accidental outbirths of spontaneous variation in the functional activity of the excessively instable human brain, which the outer environment simply confirms or refutes, adopts or rejects, preserves or destroys—selects, in short . . .* (1956: 245–246)

James, having read his Darwin carefully, clearly rejected instructionist accounts of Mind and memory. James's masterpiece, *The Principles of Psychology* (1890/1981), was one long argument for the application of selectionist principles to problems of Mind. Thus, much of the present work will be a review of James's accomplishments in the field of Mental Darwinism as well as an extension of his arguments into the realm of memory and personal consciousness.

I turn now to a summary of reasons why the selectionist approach might be useful in the area of memory theory.

WHY A SELECTIONIST ACCOUNT OF MEMORY IS APPROPRIATE

First, as mentioned above, selection involves a reduction in number of states, traits, entities, or elements from an initial plenitude of possible alternatives. As Ward (1989) has pointed out in his discussion of the similarities between selection and information theory, the information content of a signal from a given source depends on the extent to which that signal represents a reduction in the total array of signals possible, in principle, from that source. Information transmission, in other words, involves a process of noise reduction and signal enhancement. Because selection effects a similar reduction in the number of possible alternatives, it necessarily produces information content in the organism undergoing the selective action—one trait is preserved while others are not. To the extent, therefore, that memory involves the acquisition and retention of information, it must be a selectionist process. Darwinian selection seems to be the preferred way nature affects information content in an organism.

A second reason why selectionist models might be useful for memory theory is that selection requires persistence of traits through time, that is, cumulative retention of a selected group of traits. Natural selection leads to the accumulation of beneficial variations in the organisms that survive the weeding-out process. One of the things all memory theories seek to explain is the retention or persistence of information through time. Selection promotes preservation, accumulation, and retention of selective traits and disallows continuous and random change in traits subject to selection.

In the realm of the mental, cumulative development of mental content or knowledge must be the result of a process that generates candidates for inclusion into the knowledge base from the knowledge base itself. Thus, the stuff from which new "memories" are created will be related to the memory images already in the knowledge base. From a neurophysiological point of view that relation could be structural or connectional (between, for example, neuronal elements). From a cognitive point of view, that relation could be causal or semantic or meaning-based. The easiest way to acquire these "related" memories is, as already mentioned, to generate them from the current base, using current material to generate new variants on that material. This memory-generation process will involve a proliferation of images that are—at least initially—semantic or causal or structural copies and associates of the original images. Once the new variants are created, a selection process will intervene to preserve those most "fit" and these will be added to the base. Next, a new generation cycle occurs starting from the new, more fit knowledge base, and so on. This cycle of generation, selection, and retention, of course, leads to an evolutionary developmental process whereby the organism or system undergoing selection achieves a better and better fit to the environment that acts as the selecting agent. In the realm of the Mind, I will argue that this process builds an autobiographical memory system when the images/memories involved are self-related or personal.

A third reason for favoring selectionist models in memory theory has to do with the putative adaptive functions of memory itself. If memory did not exist, each encounter in our lives would be completely novel and therefore potentially dangerous. Nature can, in effect, anticipate every possible relevant future for a given organism by providing for a generator mechanism as discussed above—a mechanism that produces an array of traits (memories) that can "match" any possible relevant contingency in the environment. An organism that comes equipped with the ability to entertain various possibilities and then to match the most appropriate possibility with current environmental contingencies clearly has the advantage over an organism that doesn't come so equipped. An "instructionist" organism, for example, would have to wait for changes in the environment before it could pick up information about those changes. A selectionist system comes equipped with the relevant information, and when a match occurs between the information and the environment, the information is retained. Neither the instructionist system nor the selectionist system knows beforehand which set of environmental contingencies will be relevant (say, for survival). The selectionist system adjusts for this blindness by producing as many variations as it can to meet unknown contingencies. The instructionist system, on the other hand, offers no protection against unforeseen contingencies.

A fourth reason for exploring selectionist models in relation to memory is related to the selectionist claim that memory operates with a kind of generate-test-exit processing cycle. While it may seem pointless to generate an excess of mental representations, most of which will not meet the test condition and thus will need to be suppressed or inhibited, this excess generation may actually be required for memory retrieval. Consider the fact that most of the generated representations will be, as mentioned above, associates of the target stimulus or representation. These associates will, in effect, assist in establishing the necessary contextual information within which retrieval can proceed. Context is known to enhance retrieval, recognition, and perceptual identification. Generating associates to a cue or target helps to provide context.

A fifth reason for considering selectionist approaches to memory is that neurobiological data suggest that some areas of the brain itself develop in a selectionist manner; that is, there is an initial overproduction of neurons and neuronal elements which then undergo a process of selection and elimination via pruning (see O'Leary, 1992, for review, and Rakic et al., 1986). Reviewing neuropsychological studies on brain systems implicated in memory, Squire (1994) has suggested that consolidation of memories in humans and other gradual changes in human memory may involve selective elimination of memories. Interactions between hippocampal/medial temporal and cortical sites determine which memories get eliminated. It is important to point out, however, that selectionist mechanisms are not the only ones that operate in brain development. Nevertheless, significant aspects of brain functioning appear to rely on selectionist mechanisms. To the extent, furthermore, that both brain mechanisms and selectionist systems rely on inhibitory mechanisms, such as lateral inhibition to accomplish their tasks, they are consistent with one another. Whenever possible, cognitive models should be informed by relevant neurological constraints on the system in question; selectionist pruning and inhibition appear to constitute such a constraint.

A sixth reason for preferring selectionist models in memory theory is that they are consistent with the claim that the organism will come equipped with rich internal structure (e.g., an immune system that can recognize any potential antigen or a linguistic system that can recognize any potential sentence licensed by the grammar). Since selection requires a rich array of traits that persist through time, it apparently requires rich internal structure in the organism possessing those traits—at least for complex animals like humans. Those traits, in fact, constitute the structure of the organism. Empirical work in several branches of the neural and cognitive sciences strongly suggests that humans do, in fact, come equipped with rich innate brain and cognitive structure (Barkow et al., 1992; Chomsky, 1988; Mehler & Fox, 1985).

A seventh reason for preferring a selectionist model of memory is that selection and cumulative retention appear to promote historical lineages or clear lines of descent or connection across time epochs. This lineage effect comes out of the amplification condition—the condition that new variants must be generated from those variants that survived a previous selection cycle. Each new processing cycle, in other words, begins at a higher baseline than the previous cycle. Thus, the new variants are hierarchically and historically related to the previously selected variants.

One of the most striking characteristics of autobiographical memory in humans is its developmental or historical "feel." Autobiographical memory evolves, changes with time, and, yet, like a living organism, retains its essential characteristics. I call this feature of autobiographical memory "historicity." Each person is made up of a unique history and it might be said that that history constitutes an important part of that person's identity. We would like an account of why memory should be so closely intertwined with a personal consciousness in human beings. A complete account will involve some understanding of the nature of consciousness itself. But, as we have seen, blind variation and selective, cumulative retention result in the gradual emergence of complexity, lineages, and individuals (Hull, 1980). If selection occurs at the level of mental representations, then the pool of these representations will begin to structure themselves into a dynamic, self-organizing system—presumably something like an autobiographical memory system. Such a system would then undergo continual historical or evolutionary development. Personal consciousness, I will argue later, would emerge from the dynamics of such a system.

A final reason for investigating selectionist models of memory is that these models all assume some kind of generative potency in the system being studied. In the realm of the Mind, generative potency probably involves the overproduction of mental representations and images. Memory has always been linked (by philosophers, poets, writers, and others) to imagination and creativity. Selectionist approaches may hold the key to an understanding of the creativity of both memory and imagination and their mutual interdependence. Just as nature in its profligacy generates a huge number of forms or possibilities on the chance that some range of these will match a given ecological niche, so too, the generator in a selectionist memory system would produce a range of forms (memories). Some few of these memory images would, by chance, match some informational features of the external world or from another cognitive system. This overproduction of memory images constitutes the variability the memory system works with to produce lineages of cumulated memory episodes or images. If other branches of biology are any guide (e.g., the immune system, see Jerne, 1985) we would expect the number of generated images or forms to be quite large. In the immune system, for example, there is a vast repertoire of moleclues that can produce antibodies that can recognize any possible antigen. Antibodies are manufactured by lymphocytes that course through the blood, bone marrow, immune system glands (e.g., lymph nodes and thymus) and other sites. There are approximately 10^{12} lymphocytes. B-lymphocytes produce antibody molecules, while T-lymphocytes regulate production of antibodies. Human B-lymphocytes are capable of producing over 10 million different varieties of antibodies. This repertoire of antibodies can recognize and respond to any molecule existing in the world (antigens). An antigen binds to or selects certain of these antibodies, which are then amplified by a process of clonal multiplication. Once the antigen is bound, several varieties of immune system molecules can then be activated to destroy the invader. Thus, the immune system is characterized by an initial diversity (a repertoire of antibody molecules) that can potentially meet (via selection and amplification) any challenge.

In the case of memory and mental life in humans, the central nervous system (CNS) constitutes the ultimate source of its generativity. "The basic fact is that the nervous system contains a vast number of nerve cells and fibers, perhaps ten thousand million neurons in the human cerebral cortex and at least ten thousand million in the cerebellum. Moreover, each cell makes a great number of connections: there are up to 60,000 synaptic points on a large cortical neuron" (Young, 1979: 802).

Changeux (1985), describing the cell-count work of Powell and colleagues, noted that there are about 146,000 neurons per square millimeter of cortical surface in all mammalian species. The cerebral cortex in humans has an area of about 2,200 square centimeters. Therefore, the cortex alone contains at least 30 billion neurons. Recall, furthermore, that there are thousands of possible synaptic points on a large cortical neuron and that each synaptic connection may handle a variety of neurochemical transmitters and modulators for signaling purposes. Changeux estimates that there must be about 10^{15} synapses in the human cortex alone.

Thus, the generative power of the nervous system is clearly enormous. It is this generative power that makes the memory system a source of great variability and creativity. What about the other conditions (such as variability, a selection process, and cumulative retention) for a selectionist system? Does memory satisfy these conditions as well?

HUMAN MEMORY SATISFIES ALL THEORETICAL CONDITIONS FOR A SELECTIONIST SYSTEM

Variability

There is huge variability in the memory system, manifesting itself in a vast array of memory types. All of these memory types seem to be ecologically real things. They are not simply figments of the theorists' classificatory zeal. People with selective brain lesions, for example, may lose one class of these memories but not others. Brewer and Pani (1983) discussed six types of memory: personal, semantic, perceptual, motor skill, cognitive skill, and linguistic skill. Tulving (1983) distinguished between semantic memory and episodic memory. Schacter distinguished between explicit and implicit memory (see recent review in Schacter, 1996). Schacter's distinction complements that of Jacoby's intentional versus incidental memory, but Jacoby and colleagues emphasize the importance of cognitive attributions about the past when explaining dissociations in memory performance (Jacoby, et al., 1989). When Cermak (1987) reviewed the neuropsychological literature on memory dissociations, he found that memory could be divided into procedural versus declarative varieties.

Declarative memories are further subdivided into Tulving's (1983) episodic and semantic variants, and episodic memory is further subdivided into aware versus unaware memory. Baddeley (1986) has studied the structure and functioning of working memory, and divides it into a central executive component, a visuospatial scratch pad, and a phonological loop. Magnussen and Dyrnes (1994) presented evidence for a high-fidelity perceptual long-term memory. Nigro and Neisser (1983) presented evidence for observer versus field memories. Brown and Kulik (1977) coined the term "flashbulb memories" to describe vivid memories related to publicly important events such as the Challenger disaster. Autobiographical memory has been studied intensively in recent years (see, for example, Neimeyer & Metzler, 1994). Childhood memories can be phenomenologically distinguished from adult memories as can perceived versus imagined autobiographical events (Johnson et al., 1988). Richardson-Klavehn and Bjork (1988) distinguished between explicit intentional, explicit involuntary, and implicit forms of memory, which could be probed experimentally either directly or indirectly. Other investigators have studied state-specific memories, sensory memories (as in phantom limb pain), emotional memories, olfactory memories, body memories, collective and cultural memories, dream related-memories, prospective memories, spontaneous memories, and so on.

Now each memory type exhibits its own variability. Casey (1987) provided a thoroughgoing phenomenological analysis of the variable eidetic traits of different types of memory. He divided intentional types of remembering into an act phase and an object phase. He distinguished four types of remembering "acts": remembering-simpliciter, remembering-that, remembering-how and remembering-to. Likewise, there are four dimensions under which a memory appears in the object phase: A memory can vary in its specific content, its frame, its aura, and its modes of givenness. Under the last category, Casey discusses the clarity of the memory, its informational density, its texture, and its directness. In addition to the variability associated with the act and object phases, each memory displays a series of what Casey calls eidetic traits, or the extent to which the memory image responds to "inspection". These traits include searchability, display, encapsulement, expansion, persistence, pastness, actuality, and virtuality. So, for example, one act of remembering simpliciter will produce an image with specific content and frame and be characterized by a given emotional or semantic aura as well as informational density and the like. This same memory image will also vary in the extent to which it is searchable, encapsulated from other images, able to be expanded, real or actual, and so on. Clearly then, memory displays enormous variability both in the processes that give rise to memories and in the variety and reference of the memory images themselves. It thus satisfies the first condition for a selectionist system.

A Selection Process

Memory, I will argue later, uses certain attentional mechanisms as its selection process. These mechanisms depend crucially on inhibitory control of interference effects. The neuropsychological basis of this selection process is linked primarily to frontal lobe systems.

Cumulative Retention or Amplification

In one sense, the store of memory images is the amplification process, or at least its result. Copies of a selected target image are retained in memory. If autobiographical memory mechanisms obey selectionist operating principles, then we would predict that autobiographical memory would exhibit the following characteristics: hierarchical organization with more abstract schemas at the top of the hierarchy, historicity, nonveridical or biased content, variability of memory types, and so on. All of these predictions will be discussed in later chapters.

In summary, the available evidence is consistent with the claim that memory, including autobiographical memory, obeys selectionist principles. To see this more clearly, we can use Darden and Cain's recent analysis of the logical structure of selection-type theories in the cognitive and life sciences. Darden and Cain (1989) claimed that most selectionist theories were composed of the following basic elements or conditions:

1. There is a pre-existing variability or a set of individuals that vary among themselves, but which share some fundamental features.
2. These individuals undergo an interaction with the environment, which results in:
3. Some individuals that benefit after the interaction and some that suffer (in other words, there is a causal effect involved in the selection process).
4. Some individuals with specific sets of variations will be more likely to reproduce, thereby leading to the preservation of the associated variations.

5. A long-range effect of the interaction with the environment and of reproduction of selected variations, noted in number four above, is an accumulation of beneficial variations relative to a given and persistent environment.

Put more abstractly, conditions for a selectionist system are the following:

A. Preconditions
  i. A set of Y's exists and
  ii. Y's vary as to whether they have property P and
  iii. Y's are in an environment E with critical factor F.

B. Interaction
  iv. Y's, by virtue of possessing or not possessing P, interact differently with environment E and
  v. critical factor F affects the interaction such that

C. Effect
  vi. the possession of P causes Y's with P to benefit and those without P to suffer.

D. Longer-range effect
  vii. C may be followed by increased reproduction of Y's with P or reproduction of something associated with Y's.

E. Even longer-range effect
  viii. D may be followed by longer-range benefits. (Darden & Cain, 1989: 116–17)

Mental representations (including autobiographical memories) satisfy Darden and Cain's conditions for selectionist systems.

i. A set of mental representations (or images that could serve as memory representations) exist.
ii. These representations vary in their structural, referential, and causal properties (e.g., the extent to which they accurately model a given environment; the extent to which they inform behavior).
iii. They exist, along with the Mind they are part of, in an environment that influences (e.g., via stimulation of sensory receptors) their distribution and prevalence.

iv. Some of these representations possess a property P that better matches them to relevant contingencies in the environment, and therefore these representations interact differently with the environment relative to representations that do not possess P.
v. The relevant environmental contingencies affect the interaction such that
vi. The better matched representations "benefit" or are retained in the cognitive system while those not matched to the relevant environmental contingencies are not retained.
vii. The retention of the better-matched representations result in a greater number of these (copies) in the system (e.g., autobiographical memory store) as a whole.
viii. Because the better-matched representations are now more prevalent, the system will achieve a progressively better fit with the environment (assuming a relatively stable environment) when it undergoes another selection cycle.

Lewontin's (1982) reduction of selectionist theory to the three processes of variation, reproduction, and heritability is congruent with Darden and Cain's conclusions above. The philosophers Sober (1984) and Dennett (1995) argued that the basic processes of variation, reproduction, and heritability could be characterized as a simple algorithm, which includes causal elements. Whenever the algorithm is consistently applied, evolution occurs in whatever domain the algorithm applies, including the mental. If Sober and Dennett are right, then applying selectionist principles to mental processes should not be controversial. In the simplest case, selectionist processes might be usefully characterized in the following manner:

There is X selection for Y's that have some property P iff:

1. Y's vary with respect to whether they have P
2. There is some common causal influence on Y's (presumably the environment) that makes it the case that
3. Ys' possession of P is a possible causal factor in the survival and reproduction of the relevant organisms

For purposes of memory theory, when "Y's," or mental representations, interact over time with a given environment, and the interaction results in differential reproduction rates of these representations, then we can expect an accumulation of "benefits" in the form of survival over the long term of these items.

In summary, memory satisfies all of the abstract conditions imposed by selectionist theorists for a selectionist system. There appears to be no good reason to deny that selectionist principles might obtain at the mental level. Donald T. Campbell (1960, 1990) has been the most consistent defender of the idea that mental development and knowledge acquisition occur via a selectionist process. He argued that (1) A blind variation-and-selective-retention process is fundamental to all inductive achievements, to all genuine increases in knowledge, to all increases in fit of system to environment; and (2) in such a (blind variation and selective retention) process there are three essentials: (a) mechanisms for introducing variation; (b) consistent selection processes; and (c) mechanisms for preserving and/or propagating the selected variations. I am simply extending Campbell's arguments to the case of memory itself.

OBJECTIONS TO SELECTIONIST ACCOUNTS OF MENTAL EVOLUTION

Critics of selectionist approaches to mental phenomena argue that problem solving and learning cannot be reduced to blind variation and selective retention, since blind variation implies an unconstrained search space. But this objection, I think, misunderstands selectionist principles. Blind variation does not imply random variability or a totally unconstrained search space. Variation is always generated relative to a given and relevant baseline. The search space in the generation of new variants for a given species, for example, is constrained by the genetic envelope of that species. Insofar as constraints on the search space are useful and therefore adapted to the problem at hand, the fit of these constraints to the problem at hand must be explained. Like other cases of fit or adaptation, the evolution of constraints around a problem space probably occurred via an original blind-variation and selective-retention process. Standard learning procedures, furthermore, do not explain the emergence of new knowledge or the solution to the problem at hand. That solution is likely to have come about through trial and error problem solving; that is, a circumscribed selectionist process wherein a large number of possibilities are produced and one solution is selected.

Another objection to a selectionist approach to cognition is that cognition appears to be "directional" or purposeful, whereas blind variation and selective retention are not. If knowledge accumulation or conceptual evolution is fundamentally like the biological evolution of organisms, then conceptual evolution must not have any inherent suprahistorical goal, purpose, or directionality. Although there appears to be "progress" in human affairs, particularly in science, we cannot hold, if we adopt selectionist principles, that progress is toward some ultimate, invariant truth. Rather, progress is always relative to a given ecological niche. Selection does not preclude something like directionality in the evolution of mental phenomena; indeed, that is what cumulative retention and evolution are. On the other hand, selection does imply, that purpose will always be linked to the pragmatic needs of the organism itself. Selectionist mechanisms provide no guarantee that the cognitive system will be designed to track "truth." Rather, the truth will be tracked if it serves the pragmatic needs of the organism.

If selectionist mechanisms cannot be relied on to produce a cognitive system that tracks the truth about a given environment, then how can we claim that blind variation and selective retention mental processes produce knowledge? If we hold knowledge to be a progressively better fit of the information content in a given cognitive system to the environment of that cognitive system, then I believe selectionist mechanisms provide an adequate account of the origin of that content. If, on the other hand, we hold knowledge to be the ability to generate true beliefs, then we have to qualify the claims made in the name of selectionist approaches. True beliefs will be generated by a selectionist system only when such generation will not interfere with—or be required by—the pragmatic needs of the organism. Although some apparently believe that true beliefs are necessary to the survival of an organism, this is probably not correct. After all, there are many organisms that survive quite well without generating what we would call true beliefs about the environment. Indeed, we have abundant evidence from history books that true beliefs are not required by human beings in order to survive and even prosper. Because true beliefs are not always required for the organism, a selectionist cognitive system will likely not always produce them.

As Stich (1991) and other philosophers (including James and Bergson) in the pragmatist tradition have argued, cognitive processes should not be thought of as devices for generating truths. Rather, they should be thought of as something like tools or practices, which are to be used for very concrete and practical purposes. If a given cognitive system leads to or is useful in achieving ends that a given organism finds intrinsically useful, then that cognitive system is successful by selectionist or pragmatic evaluative criteria. This last point leads to a final objection to adoption of selectionist models for cognition. That objection suggests that such models lead inevitably into the morass of relativism.

Given the pragmatic cognitive-evaluative criteria I just described as flowing from a selectionist account, it is easy to see why the worry about relativism arises. If the criterion we invoke in cognitive evaluation is usefulness, then one person's version of the truth is just as valid as another's, as long as both are equally useful for each of the individuals involved. But this criterion does not imply that "anything goes" for anyone. Usefulness is an extremely stringent constraint. As Stich (1991) points out, pragmatic evaluation criteria in the area of technology or tools does not lead to a nihilistic free-for-all in the use of new tools or technologies.

In summary, selectionist models of Mind and memory posit huge initial variability and selective reduction of that variability via selective retention, which leads to increasing structure, information content, or knowledge acquisition in the organism, plus the accumulation of beneficial variations or lineages and histories. Human memory appears to satisfy all the theoretical conditions it must to be considered a possible selectionist system. Our next task is to show that human memory empirically operates via selectionist mechanisms. Before summarizing the relevant evidence, I will need to summarize some arguments against instructionist accounts of memory. With instructionist objections answered, I'll then turn to a summary of Bergson's memory theory as it clearly depicts how selectionist remembering might actually work.

CHAPTER 2

Limitations of the Instructivist Account of Memory

THE INSTRUCTIVIST ACCOUNT OF MEMORY

The instructionist position on memory is what I would call the standard or common-sense view of how memory works. Instructionist accounts of memory assert that memory acts as a storage receptacle for information that is somehow used by the Mind when thinking, problem solving, decision making, and so on. More fundamentally, memory supports associative learning. When I repeatedly encounter stimulus *x* in the presence of *y*, I store that association in memory so that when I later encounter *x* I can then point to y as its associate. Perhaps I could even predict the occurrence of *y* given *x*. Memory is required for associative learning because the stimuli need to be retained long enough for the association to be made—for *x* and y to be connected.

According to this common-sense view, memory promotes knowledge acquisition because it allows us to retain information that can be used in thinking. But how is retention possible? The instructionist view claims that the Mind soaks up impressions from the environment and then stores impressions that somehow represent the to-be-remembered information. The brain's cellular structure is modified in such a way as to be able (when given appropriate activation) to produce firing or discharge patterns that will represent or stand for the remembered objects/events. Memory, in this view, is a neurophysiological process involving the formation of a mnemonic trace or imprint in the brain (possibly via a strengthening of the synapses, which allow two or more neurons to communicate with one another). This imprint, then, somehow represents or faithfully preserves the to-be-remembered event. Memory is more or less like a camera: It faithfully and neutrally records present experience as it flows into the past and then stores that experience in a neutral receptacle for later recall. Both the folk and the standard scientific views of memory conceive memory as a process that accurately stores information for later recall. How else could two people who have experienced the same event agree on what actually occurred? Each person's mind must have accurately recorded the essentials of the event in question so that with later recall the two witnesses' testimony are in agreement. For all its flaws, the traditional view argues, memory is, in fact, a relatively accurate guide to the past. This standard view of memory is obviously an instructionist position. The environment instructs the memory system by imprinting information into it.

PROBLEMS WITH THE INSTRUCTIVIST ACCOUNT

While something like this standard view of memory must be correct, there are many problems with it. I shall discuss only four of them: (1) the problem of the historical character of memory; (2) the problem of the veridicality of memory; (3) memory's constructive and reconstructive abilities; and (4) the problem of the personal nature of memory and remembering.

Historicity of Memory

Any theory of memory has to provide an account of how memory comes to exhibit its historical or developmental feel. The system exhibits continuity over time. Its present is causally connected to its past, and its past is causally connected to its future and so on. Like a living being, memory exhibits a developmental pattern and progressive differentiation in its structure. Memory evolves.

A selectionist account has no trouble explaining memory's historical character. There is continuity between past and future memories because new memories are created, in part, out of old memories. Selectionist mechanisms imply that current memories are derived from a combination of old memories, a cue from the present environment and associates of that cue. The "old memories" and the associates are generated from the current stock of memories in response to the cue. Thus, new memories are necessarily related to or tied to old memories and continuity is ensured. Development occurs because the array of associates presented for selection is always generated from the current (baseline) stock of memories, and this baseline keeps growing or moving upward when new selections are added to the store. The memory store, in other words, is cumulative. Cumulative retention, in turn, implies that memory will exhibit hierarchical structure, with more complex, abstract, or differentiated memories at the top of the hierarchy.

The instructionist account of memory has no unforced way of explaining the historical character of memory. If memory involves only the transfer of information from the environment into the system, then it is difficult to see how continuity is established between old and new memories. Theoretically, the instructionist position does not even require the preservation of past memories. If the essence of knowledge is the transfer of information from the environment into the organism, old memories are not strictly required. Transfer can occur into an empty system. Indeed, many instructionist accounts argue for the blank-slate view of memory. Even if the system were not empty continuity between old and new memories would not be likely to occur in an instructivist system. What drives change in an instructivist system is input from externally derived impressions. There is no reason to suppose that continuity between old and new impressions would occur. Even if externally derived impressions exhibited some kind of regularity in their impact on the memory system, there is no reason to suppose that that impact would lead to cumulative content in the system. The mere accretion of similar impressions or images in a memory system does not necessarily translate into a bootstrapping process. Instead, it leads to a large collection of disconnected images. For development or evolution or increasing fit of organism to environment, the system needs selective retention of useful "images" only. The gradual accumulation of useful images relative to some environment confers continuity of the system's images. They are connected by virtue of the fact that each new set of images is a new and improved edition of the old set of images. Thus, natural selection leads to increasingly better adaptations to a given environment.

Veridical Memories?

Modern researchers of memory who rely solely on an instructionist approach to memory are forced to hold what appear to be contradictory positions on the accuracy of memory. On the one hand, they contend that memory accurately retains information from the past; on the other hand, they point out that veridical memories, in the sense of copies of original events, probably never occur. The problem of veridical memory storage was well known even to the ancients. The idea that memory impressions were somehow stored in the soul raised the problem of how that storage occurred. One possibility (still debated today!) was that a representation or image of the to-be-remembered event was stored in the soul and then used to later recall the event. The problems with this view were immediately apparent. Aristotle put it this way in *(On Memory and Reminiscence):*

> *But if memory really occurs in this way, does one remember the image or that from which the image came to be? For if the former, we would be remembering nothing absent (past); if the latter, how can we, while perceiving this, remember the absent thing, which we are not perceiving? If the image is like an imprint or trace in us, why should the perception of this very thing be the memory of something else and not simply of itself? For in exercising memory one contemplates the image and this is what one perceives. How then does one remember what is not present? For this would imply that one could also see and hear what is not present.* (Quoted in Krell, 1990: 17)

If the past is past, then it is not present—it is absent. But if it is not present, what is it we remember (since remembering occurs only in the present)? If what we remember is an image or representation or copy of the original event, then why do we not take the copy for the original? If the copy is not really a copy of the original event, then the veridicality of memory needs to be explained. If it is true that we, in fact, remember via an image or copy/representation of the original event, then what we remember is, in fact, not absent or past (since the image is present in consciousness) so we really are not remembering at all. Nothing from the past, then, is in consciousness. The view, therefore, that we remember only in the present is incorrect or the representational view of remembering is incorrect, or both are incorrect. Despite the obvious problems with the representational view of memory, it was popular even hundreds of years after Aristotle. Plotinus, for example, felt obliged to reiterate old objections to the theory.

If memory were a simple record of past experience, all that we would have to do to retrieve past experience would be to read the memories right off the record. We would have no need of engaging in the mental effort we call recollection.

The representational-instructionist view of memory is still what I would call the modern standard view of the nature of memory. It and its related "trace theory" of how the brain "stores" memories constitute the background assumptions of much of modern research into memory. Several philosophers of memory (Malcolm, 1977; Krell, 1990) in both the analytic and non-analytic traditions have presented arguments against the standard view and I won't reiterate them all here. Bergson, however, added a new twist to the arguments against the instructionist position, and it is worth briefly summarizing it as it informs his alternative selectionist approach.

According to Bergson (see discussion in Deleuze, 1991, pp. 58–59), if memory concerns the persistence of the past into the present, as the modern instructionist-empiricist tradition asserts, then one needs to assume that time is linear and flows from past to present to future. The present needs to continually become the past. But the present cannot become the past unless the past in some sense exists, that is, has some ontological status. Each current present must be capable of pastness. For the present to pass, to flow, the past must be part of, or available to, the present. At what point do the two temporal dimensions collapse into one another? Bergson suggested that the metaphor underlying all of these storage theories of memory and the past is a spatial metaphor. If we treat time spatially, then it follows that memories are also spatial entities and may be "stored" away somewhere in a container for later recall. Bergson suggested dispensing altogether with the spatial notion of time and memory (pointing out that physics and experience provide no basis for the spatial metaphor anyway). In its place he presented the idea of duration—the lived experience of time which is composed of a heterogeneity of non-spatial qualitative differences—a multiplicity of memory images. We humans are always steeped in a virtual infinity of these memory images. We do not experience this multitude of images because the brain shields us from them. The brain is primarily an inhibitory device. It performs its selective function by inhibiting irrelevant images and preserving only relevant images in consciousness. We can glimpse the larger realm of duration when the brain relaxes its inhibitory vigilance, as in the dream state, but usually we maintain a pragmatic focus on daily tasks, which prevents us from being overwhelmed with the stream of memory images Bergson called duration.

Thus, Bergson appears to privilege the past (duration) over the present when it comes to theorizing about memory. The relation of the "past" to remembering, however, is not perfectly clear. If we take "pastness" alone as crucial for memory experience, we find that the relation between inferences to the past and the subjective, personal experience of remembering are tenuous—the two processes can be dissociated. Even when the source of a piece of information is judged to be the past, conscious recollection may still not emerge. I can, for example, realize that I am being influenced by the past, but still not experience remembering. I can also experience the sense of familiarity that usually comes with normal remembering but still not "remember," as in the déjà vu and tip of the tongue experiences. Amnesics can refer to the past but cannot recollect the past. We can speak about the past with no accompanying remembering experience. If asked if I ever went to school, for example, I simply say yes. I do not need to make an inference to the past to recollect the fact that I went to school. In the neurology clinic, one sees the opposite case as well—a patient who believes he is remembering something when he is not, as in confabulation. The experience in these cases cannot have been based on an inference to the past, since no such past ever existed.

Adopting Bergson's position on the past as nonspatial may not solve the problem of the relation between remembering and pastness, but settling for a spatially-dependent view of "pastness" will not work either. A provisional solution may involve an agnostic position about the ontological nature of the past as well as an evolutionarily inspired notion of memory. Memory encodes a notion of pastness because it undergoes change, cumulative development, evolution. It does so because of its selectionist operations and its relation to a personal consciousness. As William James pointed out, what makes remembering cognitively unique is the reference to the past and its reference to a self. Remembering involves a pointing to the "past" in the realm of a personal consciousness—it is inextricably tied to self and identity, though not reducible to these. References to pastness are always relative to development in this personal consciousness. When we remember something we know (a) that it occurred, (b) that it occurred in the past, and (c) that it bears some identifiable relation to the self. Memory yields knowledge about the past for persons.

Constructive Activity in Memory

Instructionist accounts of memory would have us believe that memory is an essentially passive storage receptacle with minimal specialization and structure. Associative structure in memory models usually implies something like the following: Items that are in some way (usually semantically) related are stored close to one another, while items that are only distantly related to one another are stored farther apart. Even this minimal kind of associative structure, however, implies that the system performs computations on its constituent representations in order to "decide" where to store incoming items.

Studies of memory disorders (especially in brain-damaged populations), furthermore, suggest that memory is a richly structured system (Cermak, 1987), which has computational powers that far surpass those attributed to associatively structured systems. There are patients, for example, who lose their episodic memories (memories for personally experienced events) but not their semantic memories (memories for language-related meanings) and vice versa. Similarly, procedural (habit) memories may be impaired in Parkinson's disease but not declarative memory (memories that can be declared verbally), while the opposite is the case for patients with Korsakoff syndrome (Oscar-Berman, McNamara, & Freedman, 1991). The development, furthermore, of various memory-related structures appears to follow a predetermined biological plan, in that major spurts in memory ability and in memory subtypes (e.g., episodic versus semantic memory) take place at roughly the same ages and in roughly the same sequence in all normal children. Development of memory, therefore, is not tied to quantity or quality of sensory impressions available to the child. What any given child remembers is always more than (if the child remembers anything at all) and better organized than, what the child was initially exposed to. The child's memory is creative. Each act of remembering is, for better or worse, a creative achievement. A substantial amount of computation is accomplished in each act of remembering.

Other evidence suggests that the contents of memory are subject to substantial computational transformations during each instance of remembering. The literature on memory distortions and "creative remembering" is now quite large (see Ross &Buehler, 1994, and Schacter, 1996, for recent reviews). There is now no question but that remembering often involves considerable reconstructive operations. The things that get "preserved" for later recall often reflect the conscious purposes of the rememberer rather than "the true version" of what really happened. Similarly, what we recall is often constrained by what we believe to be consistent with a given identity status (Neimeyer & Metzler, 1994). Memories are often subject to a schematization or "smoothing" process, which obeys the rules for the generation of well-formed narratives (Bruner, 1994). Unless the "memory" can be assimilated into a story it will not be retained. We remember some things better than others, suggesting that memory exhibits preferences in the types of information it processes and is pre-attuned to record certain types of information. Nor do the contents of memory reflect only salient "fitness"—related features of the local environment. We remember things that go beyond the immediate experiences of our daily lives. The products of human consciousness and of human memory are, therefore, somewhat arbitrary with respect to the features of any given local environment.

Human behavior, in other words, is not a mere reflection of past experience. In order for memory to support the myriad types of creative behaviors humans exhibit, it must be generative and it must involve some imaginative reconstruction of past experience. Much of what we call remembering involves the active cognitive reconstruction of what we surmised took place in some given region of the past. Memory, like the human Mind itself, is therefore best characterized as a highly specialized and intricately structured physiological system, which preferentially handles certain varieties of information and not others. Logically, it must be that way, first, because all remembering is in service to some personal consciousness and, second, because human behavior itself is intrinsically creative.

Remembering yields special types of knowledge that impact on experience, not the least of which is knowledge about when the rememberer is being influenced by the past. When people realize that they are being influenced by the past, they can choose not to let the past be the sole determinant of their choices (Jacoby et al., 1989). If escape from the past were not possible, behavior would involve simple repetition of previously emitted acts. Thus, we can say that escape from the past via conscious recollection supports the development of creativity in human choices and decisions. Memory allows us to know when we are being influenced by the past, and we can use that knowledge to escape the influence of the past. Paradoxically, then the function of memory is not merely to preserve the past; rather, the function of memory is also to allow us, as personal individuated beings, to escape the past.

Creative reconstruction of past events in memory has an important impact on subjective experience. As past events are imaginatively reconstructed, numerous memory systems are activated and the possible interpretations of current events multiply. These multiple interpretive possibilities for any given current event, in turn, allow a person to entertain any number of alternative responses or alternative identities. Humans are free, in other words, to choose not to repeat the past because the number and variety of memory systems—as well as their ability to generate novel representations that can support creative and imaginative alternatives—confer on human beings a new kind of autonomy from the past. Without this imaginative component to mnemonic processing, we would not be able to escape the influence of past events on current behavior or experience.

The activity of the imaginative or generative component of memory brings with it a puzzling set of experiences. First, it creates what can only be called an excess of mental products or images. Second, it introduces us to an essentially invisible realm that we sometimes call the past. It is as if we experientially assign to the present all things that we can see or currently perceive, and we assign to the past and memory all those things we cannot currently perceive but can nonetheless still (somehow) experience (in memory). Memory is the experience of things unseen—not just the things once seen.

Neurological disorders sometimes allow us to catch memory in the act of creatively producing a range of possible "memories" for an individual. Korsakoff syndrome is usually associated with chronic alcoholism and thiamine deficiency. The most common lesion site involves the dorsomedial thalamic nucleus and the mammillary bodies. Both of these anatomical sites send efferents to prefrontal cortex, one of the most evolutionarily advanced areas of the cortex. The main neuropsychological deficit, however, is a severe antereograde amnesia (the inability to learn new material). Compounding a general loss of interest in anything but the present moment is an extensive retrograde amnesia (loss of old memories) so patients have no ready store of personal memories to share when inquiries concerning their past are made. When asked about their past, some patients will provide only vague generalities; others will make up fantastic stories that cannot possibly have been true. The latter strategy is known as confabulation. Patients believe they are reporting true memories and one gets the feeling when witnessing an episode of confabulation that one is present at the birth of new "memories." I once, for example, witnessed an interview between a neurologist and an amnesic patient who was clearly confabulating. The conversation went something like this:

Neurologist: *So, why are you here in the hospital?*

Patient: *Yes, I served on a destroyer in the war. (This may have been true. The patient was a World War II veteran.)*

Neurologist: *But why are you here now?*

Patient: *I was wounded in the head.*

Neurologist: *(deciding to go with the patient): Your ship was hit and you got hurt?*

Patient: *Yes, I was washed up on the Italian coastline.*

Neurologist: *Ah, yes and you were then harassed by enemy agents?*

Patient: *Yes, The Vatican and Israeli agents were out to get me.*

Neurologist: *Didn't you write up an account of your adventures?*

Patient: *Yes . . . there is a book . . .*

Clearly, this patient exhibited no awareness or insight into his memory problem or his tendency to confabulate. He later produced the most fantastic and convoluted tales concerning his World War II adventures; these tales were always elicited when he had to search his memory. The imaginative component of memory was operating without inhibition. Other patients do show some insight or some awareness that their confabulatory stories seem odd and fantastic.

Children also confabulate, often in response to queries about where they have just been or what they have just been doing. Once again we can witness memory giving birth, via "confabulation," to new memories. In this case, the process can in no way be considered pathological. Rather, it is a natural phenomenon associated with development and maturation.

Memory and Personal Identity

Instructivist accounts of memory have trouble dealing with the biased nature of recall and with the old problem of how identity emerges from memory. If memory primarily involves the transfer of information from environment into the organism, then memories should primarily and accurately reflect the environment rather than the rememberer. Similarly, if memory essentially reduces to a transfer of environmental information into the organism, then the organism's structure should reflect or track the structure of the environment. Individuals living in identical environments should have identical memories/identities. But because individuals are not mere reflections of their environments, instructivist accounts fail. Selectionist accounts, on the other hand, assume that remembering always takes place within a personal consciousness and that remembering is always in service to this personal consciousness. Memories always reflect a personal point of view, so remembering is necessarily biased. It cannot be otherwise, since identity emerges from autobiographical memory. No identity—no "sameness" of responses or behaviors through time—is possible without some function relating one behavior to another across time epochs. Memory's retentive function allows us to see patterns in our past responses and those patterns become us. Selectionist accounts of autobiographical memory suggest that continuity of self (identity) is the result of the ways in which cumulative retention of selected memories works. Selectionist mechanisms require that new memories be developed out of old memories in interaction with the current environment. Development involves a synthesis of past information with current information, so continuity of identity is ensured.

The relation between identity and memory is very intimate and somewhat obscure. I can see the world only through the lens of my particular identity and that identity was forged for me through a whole series of past experiences that have been retained in memory. When I look into my past, I believe that I perceive a common thread weaving itself through all of my experiences up to now, but the thread is quite fragile. The novelist Marguerite Yourcenar expressed it this way:

> *The landscape of my days appears to be composed, like mountainous regions, of varied materials heaped up pell mell. There I see my nature, itself composite, made up of equal parts of instinct and training. Here and there protrude the granite peaks of the inevitable, but all about is rubble from the landslides of chance. I strive to retrace my life to find in it some plan, following a vein of lead, or of gold, or the course of some subterranean stream, but such devices are only tricks of perspective in the memory. From time to time, in an encounter or an omen, or in a particular series of happenings, I think that I recognize the working of fate, but too many paths lead nowhere at all, and too many sums add up to nothing.* (1954: 24–25).

If I stop and hold audience with my memories, do I find in them the thread of my identity? If I start by trying to recall my earliest childhood memories, I come up with one or two recollections. But these recollections can't be entirely faithful reproductions of what actually occurred, since the memories consist of me seeing myself doing something (like playing with some rocks in a yard). If it is impossible for me to step outside of myself and observe the proceedings from an independent stance, then the memory must be an imaginative reconstruction of something that actually did occur. I do remember playing with those rocks, but my memory of me playing with those rocks is of a disembodied me watching the embodied me playing with those rocks. The mnemonic knowledge I have of playing with those rocks as a child is dissociated from the "memory" image I call up in association with that mnemonic knowledge. Do both of these mnemonic sources contribute to my identity? Even if I supposed that they did, what do I divine about my identity from that memory of me playing with those rocks? If I add to it a whole series of other images, other "memories" leading up to the present day, am I then able to piece together the profile of a person, a plan of some kind, a consistent meaning, a pattern of behaviors? I grew up in a family, I went to school, I met friends and I developed these friendships, I married, I had children, I fathered these children, I accomplished a few things. . . . Does all of this add up to an identity? Are they not the story of every man—not my particular self? While I can certainly ascribe some meaning or purpose to my life, it is not at all clear that that purpose can be derived from an examination of my memories. A thread of connected memories does not automatically yield a single underlying meaning. Yet where else but memory should we look when we ask the question of identity: "Who am I?" The problem is that memory answers: "You are many."

When I look into myself to examine who I am, I begin by consulting my memory, but find there a vast and phantasmagoric menagerie of images. What images become available? Perhaps a single picture or image of a person emerges but no particular memory. When I stay with that image, then more will emerge: pictures of that person I supposedly was in the past. I can see the person doing something, wearing characteristic clothes and moving in characteristic ways. I now remember certain episodes and feelings associated with those episodes. The person seems changed, more grave and serious. But now new episodes emerge and the person changes yet again. If I stay with these images, new ones emerge.

Now it becomes more difficult to say whether these new images are really memories. At some point, our memories become fantasies. Memory invents more or less plausible "memories" when we consult it to uncover the histories of ourselves and others. Memories are experienced as a series of metamorphoses of the self. There are some constants, like desires, strivings, and behavioral preferences. But can we reduce the self to a set of appetites? Of drives? If I pursue my memories further, I can call up some images of myself that involve no particular drive, desire, or striving. Indeed, these are some of my happiest memories (e.g., of wandering in the mountains and forests). But they involve a self that seems diametrically opposed to the desire-ridden self. If I pursued the recollection process further, I am sure that I would discover other selves as well. It is important to point out that these selves are creations of memory. In daily life I necessarily experience myself as an indivisible unity. Apart from pathology, so does everyone else. Whence come these multiple selves in memory, then? They can only come from memory processes. Just as the Freudian dream work of symbolization, condensation, displacement, and the like gives rise to the manifest dream content, so do memory processes produce many selves. This is not surprising given memory's creativity. One reason we moderns avoid extended periods of recollection is precisely because we find in those recollections no unified, centralized ego. We therefore attempt to ignore the evidence of the mnemonic senses in order to present a unified front, a single mask, to the world. The attempt, however, is largely unsuccessful, so the efforts to secure a single, unified identity that remains constant across all contexts is probably futile. Memory, therefore, presents to us a paradox with respect to its role in identity formation. On the one hand, memory must constitute the basis for personal identity, since it preserves a record of the story of our lives from start to finish; yet when we look into our memories—our past and formative experiences—we find material that imaginatively constructs multiple identities, which seem rich, profound, and authentic to us.

Augustine expressed something of the mystery of memory and identity when he wrote:

> *Great is the power of memory! It is something terrifying, my God, a profound and infinite multiplicity; and this is the mind, and I am this myself. What therefore, am I, my God? What is my nature? A life various, manifold, and utterly immeasurable. Behold the plains, caverns and abysses of my memory; they are filled beyond number with innumerable*

> *kinds of things, present either in their images, as in the case of all bodies or by means of their own presence; as with the arts, or in the form of some kind of notions or impressions, as with the affections of the mind which, even when the mind is not experiencing them, the memory still retains, although whatever is in the memory is also in the mind! Through all this I range in all directions and flit here and there. I dive down as deeply as I can, yet there is no limit. So great is the power of memory, so great is the power of life in man who lives mortally. (Augustine, 1979, Book 10: 214–215)*

If memory is very much like the imagination in its creative trafficking in images, episodes, identities, and stories, then memory, like the imagination, must be a primary source for the "stuff" that constitutes personal experience. Despite memory's immersion in fantasy, we nevertheless impute to it the qualities of accuracy, truth, veridicality, and facticity. There is something about the subjective experience of remembering that compels belief. Nature, apparently, wanted us to give special weight to the knowledge yielded up to us by the memory faculty. An instructionist account of memory neglects these creative aspects of remembering. We turn now to the selectionist alternative, taking Bergson's reflections on memory as a starting point.

CHAPTER 3

Bergson's Memory Theory

The two philosophers who were the first to see the importance of Darwinian selectionism for the study of memory were William James and Henri Bergson. James's efforts will be discussed in the next two chapters. Here I focus on Bergson's selectionist account of remembering.

BERGSON

Henri Bergson (1859–1941) was one of France's most prominent turn-of-the-century philosophers interested in problems of psychology. The range of his psychological interests was astonishing and included consciousness, perception, habits, memory, the aphasias, dreams, laughter, intuition, time perception, brain functions, multiple personality disorder, the déjà vu experience and much else besides. His early works *Time and Free Will* (1912), *Matter and Memory* (1888), and *Creative Evolution* (1907) announced a new and strikingly original approach to Mind and human nature, and influenced a number of psychologists then active in Europe and the United States. James, for example, described the experience of reading Bergson in the following way: "Open Bergson and new horizons loom on every page; it is like the breath of the morning and the song of the birds" (1909: 270). Freud also was impressed with Bergson's works. When the great Boston neurologist James Jackson Putnam wrote to Freud in 1910, urging him to consider Bergson's views on memory and the brain, Freud replied, "I make bold to reply that from the beginning I have agreed with Bergson's view of the matter" (quoted in Hale, 1971: 96–97). Bergson was one of the first philosophers (indeed, one of the first scholars) to use data from neuropsychological dissociations to illuminate problems in the philosophy of Mind and cognition, and certainly the first theorist to produce a relatively complete selectionist account of memory. The intellectual climate within which he worked was saturated with speculations concerning the impact of the Darwinian revolution on all the other sciences. Many investigators in the human sciences were beginning to apply selectionist ideas to their particular fields. Psychologists, in particular, were intrigued. The American psychologist, James Mark Baldwin (1861–1934), was probably also influenced by Bergson since he spent a good part of his career in France when Bergson was active. He observed: "And how far the method of law called by Darwin 'natural selection' goes, what its range really is, we are now beginning to see in its varied applications in the sciences of life and mind. It seems to be—unless future investigations set positive limits to its application—a universal principle; for the intelligence itself, in its procedure of tentative experimentation, or 'trial and error,' appears to operate in accordance with it" (Baldwin, 1889: 83). Both Bergson and James studied the works of the "social Darwinist" philosopher Herbert Spencer but found them unsatisfactory.

Bergson explored the problem of memory in one of his earliest and greatest works: *Matter and Memory* (originally published as "Matière et mémoire: Essai sur la relation du corps avec l'ésprit," Paris: F. Alcan, 1888). Although Bergson's major concern in that work was to develop a solution to the mind-body problem, he also apparently felt obliged to develop a detailed theory of memory.

THE THEORY OF MEMORY

For Bergson, the primary function of memory was:

> *to evoke all those past perceptions which are analogous to the present perception, to recall to us what preceded and followed them, and so to suggest to us that decision which is the most useful. But this is not all. By allowing us to grasp in a single intuition multiple moments of duration, it frees us from the movement of the flow of things, that is to say, from the rhythm of necessity. The more of these moments memory can contract into one, the firmer is the hold which it gives to us on matter: so the memory of a living being appears indeed to measure, above all, its powers of action*

> *upon things and to be only the intellectual reverberations of this power.* (Bergson, 1888/1988: 228).

This passage succinctly summarizes Bergson's theory of remembering. When remembering is triggered by a current perception or cue, a series of images and past perceptions become available to the rememberer. These memory images must be related to the current perception—that is, they must be similar or analogous to the current perception. In addition, all of those memory images that preceded or followed the past perceptions, which are most similar to the current perception, are also recalled. The current perception, in other words, triggers an approach of a vast range of memory images that are formally—even distantly—related to it. (This approach of the past toward the present is essentially what we have been calling the generation of memory images or associates, but Bergson calls it "multiple moments of duration.") Once this vast array of images are activated, they are used to suggest to us the decision that is most useful. Presumably, the images are used to help contextualize the search for the memory image that best matches the cue or trigger image. We will see later that Bergson claims that irrelevant images are inhibited or suppressed during the search process. The selection criterion is "usefulness," or those images which promote the autonomy of the rememberer. Memory, according to Bergson, is firmly bound up with personal freedom and autonomy. Conscious remembering lets us know when we are being influenced by the past. Once we have this information, we can choose whether or not to let the influence proceed. Without conscious recollection we would have no choice but to act out the influences of the past. Similarly, conscious recollection, allows us to temporarily withdraw from the pressures on the present, thereby giving us distance from ongoing concerns and obligations. Remembering, in sum, is crucial for freedom and autonomy from both the past and the present. It balances our investment in the present with an attachment to the past, and it confers on us a choice about when to be influenced by the past.

Activation of memory images is experienced in a single intuition as "multiple moments of duration." For Bergson, "duration" or (roughly) lived time, is juxtaposed with abstract, spatialized notions of time. Every moment of duration is novel, not repeatable, and inherently unpredictable. Abstract time, on the other hand, is composed of identical, spatialized instants. Each instant totally determines the next one in the chain. Spatial time, therefore is inherently deterministic. A single episode of (Bergsonian) remembering, on the other hand, takes us out of the realm of necessity—of being acted on as if we were simply another object in successive, incremental, and repetitious time and places us instead in multiple moments of duration—multiple possible worlds, as suggested or implied by the ensemble of past perceptions now available in memory. The memory images (and their implied possible worlds) have their own rhythm, their own movement and logic, since they are moving toward a decision. Bergson calls this movement of the past toward the present "contraction." From the point of view of present perception, however, there is an expansion of memory images in consciousness, since a region of the past is becoming available to it. To the extent that memory can "contract" the moments of duration into one moment, one decision, it increases the organism's powers of action—the organism's freedom.

In summary, in a single act of (Bergsonian) recollection, we have the following sequence of events. In an initial relaxation of inhibitory processes, a vast array of images become available. These images are used to contextualize the search process. Most of these images undergo a secondary process of inhibition to ferret out the non-useful images. One set of available images (the useful set) escape the inhibitory suppression and are allowed to inform a decision concerning action.

A modern reader might formulate the sequence of events as follows: A cue, usually some current perception or some internally generated clue or guess, triggers generation of a vast array of memory images that are, in some way (perhaps semantically), related to the cue. Once an initial set of images are triggered, a further set of images are elaborated. This second set of memory images must be causally related to the first. After all the distractor images are suppressed, a match is made between the cue and any remaining internal memory-images, and that match constitutes recognition and remembering.

For Bergson most memories/rememberings are not triggered by external cues. Rather, memory has its own rhythms and laws—its own "agenda." Memory's usefulness lies in the fact that it allows us to escape the influence of the present environment and thus confers on us a certain measure of autonomy. Bergson pointed to the phenomenon of spontaneous rememberings as evidence for memory's nondependence on external cues for its operations. Spontaneous rememberings may be the norm for human beings, and should be taken to be the primary focus of study in the memory field.

According to Bergson, every instance of a spontaneous remembering is composed of a series of phases. First, the brain relaxes its inhibitory powers, then there is a proliferation of memory images, and finally a selection or preservation of a subset of those images. The selection phase, however, depends on inhibitory processes. Thus, there is an initial phase of relaxation of inhibitory powers, which might be called the proliferation or generation phase. After the proliferation of images or representations occurs, they are entered into some kind of selection or decision-making process, which depends once again on inhibitory processes. This second phase might be called the actualization or selection phase. The proliferation phase produces an array of images or mental representations that represent past perceptions as possible states of affairs or possible worlds—Bergson's moments of duration. Next comes what Bergson calls contraction—where a distinct region of the past becomes available via inhibition of a multitude of irrelevant images. A part of this process might be called "contextualization of the cue" in service of the search to zero in on the target. As contraction proceeds, the "relevant" past edges closer and closer to the present. This contraction of the past must occur so that it can address the present, but the contraction is experienced as an expansion by present consciousness, since its repertoire of images (moments of duration) increases.

In the selection phase, various "moments of duration" (or a small set of possible worlds) will be selected from the array of images or moments thrown up by the proliferation phase. What guides selection for Bergson is the concept of "usefulness" or "fitness." But Bergsonian fitness should not be confused with classical Darwinian notions of fitness. Whatever moment best promotes freedom of action is "fit" and will be selected from the array of choices thrown up by the act of remembering. The thing chosen is what we usually call the memory *x* when we say : "I remembered *x*." Freedom of action, for Bergson, entailed some measure of autonomy from current environmental cues. This selection criterion distinguishes Bergson from other selectionist theorists. Most theorists consider the selection criterion as a matching function: Select whatever best matches current environmental information or demands. But to Bergson, the criterion is whatever promotes the autonomy of the rememberer. Autonomy is conceived as a balance between past and present, between inner and outer informational streams of thought.

The selection process for Bergson cannot be a matter of mere matching (or resonance) of a memory image with the current precept. According to Bergson, matching does not work even in the case of recognition. Rather, selection is Darwinian in the sense that what is chosen is what is used—what is left over after the other possibilities/images/alternatives are suppressed or discarded. Thus, what is chosen is not necessarily the best match or the most optimal solution to a current perception or to the problem of guiding action. The selected image or model (given the proliferation and amplification phases that produced the model) may possess properties that are only distantly related to the current perception. Since the criterion is usefulness in promoting the freedom of action of the rememberer rather than optimal fit (or matching), the image that survives the selection process will probably not be some copy of a past state of affairs or of a current perception. Nevertheless, Bergson occasionally uses the term "matching" for shorthand when describing aspects of the selection process.

The foregoing is essentially the Bergsonian theory of remembering. Bergson, however, would probably not have referred (with respect to the proliferation phase) to a generation of images. Rather, Bergson suggested that we are always steeped in a virtual infinity of images. The brain acts as a kind of inhibitory device that screens out most of these images, allowing only the useful to remain. These useful images we call the world, objects, everyday reality—and memories.

As mentioned in Chapter 1, any selectionist theory of memory must be composed of at least three processing components: a generation (or proliferation) phase, a selection phase, and a cumulative retention or amplification phase. I now want to discuss some of Bergson's suggestions concerning these phases. In order to facilitate comparison with modern memory theory, I will focus on memory representations, the process of remembering or recollection, and the process of selection. Bergson's ideas on selection were introduced above. I will briefly review them again in the context of Bergson's discussion of the problem of how memory processes might be mediated by the brain.

SPECIAL TOPICS IN THE THEORY

Representations

Recall that during an act of remembering in the proliferation phase, a huge array of memory images becomes available to the rememberer. An initial set of memory images or representations and then a secondary set, causally related to the first, is generated in response to some cue. What are these images/representations? How is a representation related to the thing it represents? Is the relation one of identity or some more limited type of correspondence? Traditional theories of memory assume that memory images/representations preserve something (a trace) of the object that they represent. But this trace theory is problematic, according to Bergson. He asks why we posit an outer image (object) and an inner image (representation), which are supposed to be about the same thing, and then go on to assume that the two images must be different? Why do we not take the trace representation for the object itself? Why do we not mistake perceptions for memories? Bergson suggests that perceptions or object images are distinguished from memories or representations in the following way. Object-images are experienced by us as being bound up in an aggregate of other object images all causally related to each other in sequential time. An object image is firmly situated in sequential time. Representations (memories), on the other hand, lie outside of the series of aggregate images. A memory representation is an object image that has been removed from the aggregate—from sequential time. Its associates have all been inhibited: "To transform its [the object image] existence into representation . . . it would be necessary not to throw more light on the object, but, on the contrary, to obscure some of its aspects, to diminish it by the greater part of itself, so that the remainder, instead of being encased in its surroundings as a thing, should detach itself from them as a picture" (Bergson, 1888/1988: 36).

A representation, in other words, is obtained by progressively eliminating aspects of an object-image—especially that object's relations with other objects. The object is extracted as figure from ground by pushing away its neighbors (via a kind of lateral inhibition). In other words, we are presented with a plenitude—an aggregate of object images. To obtain a memory representation, we inhibit portions of this aggregate until we isolate a single image from its related images in the stream of sequential time. By isolating an object from its causally related objects in the stream of sequential time, (via this kind of lateral inhibition) we remove it from the realm of necessity and radical mechanism. Thus, a memory representation is distinguished from a precept or an object-image because the former has undergone selection and is now a part of duration, while the latter remains a cog in the wheel of spatialized time.

Now memory images are also considered to be representations of absent objects or states of affairs. Memory is about the past. The past, however, survives under two distinct forms: habits and conscious recollections. Here Bergson anticipates modern distinctions between procedural memory and episodic memories. Habit involves repetition, while episodic memories involve the unique, nonrepeatable event. Recollection bears a date and is a representation as defined above. Habit is effortless and, once formed, nonrepresentational. Recollection, on the other hand, requires an act of will. It involves an active letting go of the intense attachment to the present we all experience. For Bergson, then, recollection allows us to detach ourselves from bondage to the object-world and enter into the world of duration.

Retrieval

What governs or guides retrieval? In Bergson's selectionist theory of memory, the past is the active agent that comes out to meet the present. Remembering or retrieval is an act, not reducible to anything else, by which we recover the past. The past, through a relaxation of inhibition, floods the cognitive system. An array of possibilities is presented for selection. We have already seen that, for Bergson, the selection process is constrained by the need to support autonomy and to guide behavior. But this is only a very broad constraint. "What we really need to discover is how a choice is effected among an infinite number of recollections which all resemble in some way the present perception, and why only one of them—this rather than that—emerges into the light of consciousness" (Bergson, 1888/1988: 164).

Bergson suggests that retrieval works in tandem with the proliferation/generation process. The array of items generated is expanded until the array contains a class of representations that resemble the current percept or provide a context for the current target image. This expansion of memory elements in consciousness, in turn, involves two other processes: translation and rotation. These collective mnemonic processes compose the aforementioned movement of the past toward the present in an original act of remembering. "[M]emory, laden with the whole of the past, responds to the appeal of the present state by two simultaneous movements, one of translation, by which it moves in its entirety to meet experience, thus contracting more or less, though without dividing, with a view to action; and the other of rotation upon itself, by which it turns toward the situation of the moment, presenting to it that side of itself which may prove to be the most useful" (Bergson, 1888/1988: 168–169).

Thus, for Bergson, in any act of recollection we initiate the process by relinquishing attachment to the object-world. We relax our inhibitory powers so that we are able to adopt the proper receptive attitude toward the impending approach of the whole of our past. The whole of our past undergoes a kind of contraction so that it can move toward the present. The entirety of the past is translated into a form amenable to the present. The past then undergoes a rotation so that only those portions of it most relevant to the present appear in consciousness. At this point the cognitive system is flooded with a vast array of memory images (expansion). Now the brain's inhibitory powers are tapped to begin the process of contextualization. After inhibition and suppression do their work, we are left with a narrow range of images or moments of duration, and these are experienced as veridical memories of a real past state of affairs.

Selection and Mind-Brain Relations

Bergson is at pains to point out that the selection process involves a movement of the past toward the present rather than the present calling up the past. The past comes out to meet the present perception. That essentially is what proliferation/generation implies: The past is the active agent. The present percept merely and passively matches one of the myriad possibilities presented by the past. For Bergson, the respective roles of the present and the past in the selection process speak to the ways in which memory and brain interact. A sensory stimulus or perception should not be understood as activating or as becoming a memory. Memories are not weakened versions of percepts. The contents of memory do not reflect or correspond in any simple way to the things we have perceived throughout our lifetimes. Relative to the contents of memory for any given episode, the perceptual stimulus associated with that episode will be impoverished. Thus memory cannot reflect the environment, and empiricist approaches to memory must fail. Although we have perceived a huge number of faces in our lifetime, we remember only a very few of these. We have, furthermore, never perceived, in any ordinary sense of the term, the history of a long-term cooperative relationship with a friend, yet we can provide from memory detailed sets of information and detailed memories about that relationship as it evolved over time.

Since we cannot derive the contents of memory from perceptual experience, memory cannot be reduced to perceptual traces which "persist." There are no such things as memory traces. If, on the other hand, memory could be considered a derivative species of perception, then memory traces could probably be localized in the brain, since you would need a medium to contain these traces. But Bergson argues that the brain acts merely as an inhibitory or selective device. It allows, in response to perception, certain movements and inhibits others. Perceptions issue not in memories but in movements. Memory is not a regression from present to the past but a coming forward of the past to meet the present. When the brain is injured, memories are not lost. Instead, brain injuries influence the inhibitory powers of the brain and, thus, some aspects of memory are temporarily less accessible than others, but no memories are lost.

According to Bergson's selectionist assumptions, a sensory stimulus's only role is to trigger something like amplification of an item or pre-representation that is, in some sense, already "there." In Bergson's conception, all of the past is "there" in the form of the innate repertoire or in a prepotent virtual state—ready to be generated during amplification. Never would a lesion of the brain really destroy memories. Disorders of memory associated with brain damage can only be due to two things: (1) the inability to choose the appropriate memory from among the array of available images (insufficient inhibitory power) or (2) the inability to translate memories into motor actions. Beyond these distinctions or patterns of memory breakdown, Bergson did not advance any detailed theory of memory disorders. Yet we can infer that he might have predicted a third pattern: the inability to detach from a current external or salient stimulus so that one can adopt the appropriate receptive stance to the past (see Lhermitte, 1986, for a description of just such an "environmental dependency syndrome"). This inability to escape the influence of the environmentally salient present would be considered by Bergson to be a disorder in the actualization process.

Bergson's Critique of Association Psychology

Modern information processing models of memory rely heavily on ideas originally formulated by the associationists (e.g., Locke). When Bergson began *Matter and Memory* in the cultural milieu of late nineteenth century France, Taine and Ribot were the leading French theorists on memory. According to Bergson, they promoted simplistic and mechanistic models of the Mind, while borrowing heavily from the British associationists. In the associationist view, memory is structured by sets of mental representations, which are acquired through learning. Learning is governed by the laws of association: Ideas are acquired through contiguity (in time or space) or through similarity. Memory search or retrieval is possible because the search process is not random—it can, for example, proceed systematically through semantically related ideas until it finds its match. Bergson's objection to the associationist stance with respect to memory was that it was simply uninformative or trivial. Every idea has some kind of an associate, but this tells us nothing about the mechanisms of association. Association by resemblance and contiguity surely occurs, but that fact does not explain how recollection is possible. Why, during any given act of recognition or of remembering, does a single memory emerge into consciousness? The answer to this question will tell us something about the mechanism of association and the mechanism of selection. Bergson's demands for a theory of the mechanism of association has largely been met by modern cognitive neuroscience. For modern theorists of the dynamics of associative networks, selection is largely accomplished via variations on the mechanisms of (1) lateral inhibition and (2) the search process (see, for recent examples, discussions in Grossberg, 1980; and Raaijmakers & Shiffrin, 1981). Bergson, I believe, would find such explanations congenial and impressive. Raaijmakers and Shiffrin's mathematical model of retrieval dynamics, called SAM, or search of associative memory, for example, is congruent with Bergson's arguments concerning remembering. Retrieval is assumed, under SAM, to consist of a series of search and recovery operations guided by a "retrieval plan." The plan is used to determine what cues will be utilized at each stage of the search until a match is found and the search is terminated. Retrieval search samples successive sets of images until a selection is made.

We have seen that for Bergson what determines selection is "usefulness" or, broadly speaking, adaptation, in so far as adaptation implies autonomy from salient environmental stimuli. The emphasis for the selectionists, including Bergson, is at the molar behavioral level. Indeed, except for his reluctance to postulate internal structure, B. F. Skinner may be classified as a selectionist. For Skinner, the organism is always emitting a plenitude of responses. Those responses that get preserved or selected are the ones that get reinforced. There may not be a difference in kind between

Bergson's adaptation criterion and Skinner's reinforcement criterion. Skinner simply confined himself to a more narrow range of conceptions of "usefulness" than did Bergson.

In summary, Bergson's memory theory is clearly selectionist in character. There is an initial diversity (of memory images) and a subsequent selection process from this diversity. Selection for Bergson, however, could not be a simple matching process (between percept and memory image) otherwise, cognition would only be a reflection of local environmental conditions. Memory images are selected (and therefore retained) if they are useful for increasing the possibilities for free action. We turn now to the experimental evidence for Bergson's model of remembering.

CHAPTER 4

Evidence for Selectionist Processing

TESTING BERGSON'S SELECTIONIST THEORY

In Bergson's selectionist account of memory, retrieval depends on selection, but selection, in turn, depends on inhibition of irrelevant or distractor "images." The distractor images are causally or semantically related to the retrieval cue, which can be derived internally from other cognitive systems or externally from perceptual sources. For Bergson, the distractor memory images are plentiful and always there, but they are held back from consciousness by the brain's inhibitory processes. They are, however, an intimate part of the retrieval process, since they help to contextualize the retrieval search. The distractor images are generated in an initial relaxation of inhibitory processes. Once generated and allowed to contextualize the search process, these images undergo a secondary process of inhibition, when the search is narrowed down to the target memory (the image that best matches the cue). Thus, inhibition plays a crucial role in Bergson's selectionist memory theory.

Inhibition is crucial for cognitive selectionist theory in general because (1) it points to the mechanism through which selection is accomplished and (2) it raises the issue of interference effects during cognitive processing—particularly during retrieval from long-term memory. Interference effects in long-term memory would not be seen if distractors were not present. Distractors would not be present if they had not been generated or allowed by the Mind. No selectionist process would be necessary if no distractors were present. "Selection" is always from an array of alternatives, some of which are inevitably irrelevant. These irrelevant alternatives must be ignored or actively suppressed so that selection can zero in on the relevant target images. Simply letting the activation levels of irrelevant distractors relax back into subthreshold levels if not "selected" would be a slow and inefficient process.

Inhibitory mechanisms are not required by instructivist accounts of cognition. All that is needed to transfer information from environment into the brain/Mind is that the brain/Mind be (1) open to receiving the information and (2) capable of retaining or storing the information. The brain should be imprintable like wax. The environment should be able to trace patterns in this waxy substance so as to inscribe the appropriate information. Of course, the ways in which recall is accomplished in this instructivist scheme have always been obscure, but some kind of re-activation of inscribed traces was usually suggested. Inhibitory processes simply were not required.

Instructivist versions of memory required modern investigations to delve into basic neural and cognitive processes of cognition, suggesting that inhibition plays a role at every level. Inhibitory processes are ubiquitous in brain and cognitive processing (see recent reviews in Dagenbach & Carr, 1994; and Dempster & Brainerd, 1995). The reliance on inhibitory mechanisms should not be surprising. When inhibitory mechanisms are coupled with excitatory mechanisms, the brain can rapidly and selectively enhance a target signal, while suppressing background noise. By relying on dual mechanisms of activation and inhibition, which act simultaneously on both the target and the irrelevant images, rates of information processing (i.e., of extracting a signal from noise) will be significantly increased relative to systems that employ only activation or only inhibitory mechanisms. Utilization of inhibitory mechanisms in selective processing is therefore a necessary but not a sufficient indicator of selectionist systems.

Other indicators of a selectionist memory system include:

1. Evidence for the generation of multiple associates (including irrelevant distractors) to some cue
2. Use of these associates to contextualize the cue
3. Inhibition of distractors
4. Selection or matching of cue and target

5. Amplification (copying) of selected targets through the system
6. Cumulative retention of selected targets

I will briefly review the experimental and neuropsychological evidence for the first five of these selectionist processes since these, when considered together, constitute the core of Bergson's selectionist memory theory. Evidence for "cumulative retention" of selected targets is the memory system itself. There will naturally be some overlap since inhibition is key to all of the above steps.

ATTENTION AND INHIBITION

Attention is selection in the realm of the Mind. Unless an item is attended to, it is not explicitly remembered. Attention processes have been conceptualized in a number of different ways, with inhibitory processes playing a central role in most formulations. Theories of selective attention that select by enhancing the target signal, while leaving the noisy surround unchanged, are several and include those that LaBerge (1995) called the spotlight, zoom lens, resource allocation, and gain models. In the spotlight model, attention works like a spotlight or search light, illuminating targets otherwise shrouded in darkness. The zoom lens metaphor adds the elements of clarity and intensity of attentional focus to the spotlight metaphor. By allocating processing resources to attentional effort to these models, informational pickup is enhanced and we arrive at the resource allocation and gain models. Theories that postulate selection by inhibition of the surrounding area, while leaving the target area unchanged, are the filter, gate, and channel capacity models. The basic idea is that the signal can be enhanced by progressively screening out or filtering out noise. The screening process can be accomplished either via gates or limited capacity channels.

Attention theories that lean heavily on information theory tend to invoke metaphors involving channel restrictions and capacity limitations. Transmission of information is thought to occur along a channel with limited capacity in terms of amount and rate of information flow. Kahnemann (1973) described two types of models involving resource limitations: (1) those with structural limitations that predict that concurrent tasks interfere with each other because the same mechanism is required to carry out two separate operations and (2) those with capacity limitations that predict that concurrent tasks interfere with each other because the requirements of each task exceed available resources. Broadbent (1958) proposed a filter theory of attention in which only a single channel of attending occurs while information in other channels is not sampled, and so is lost or at least is reduced. Filtering implies that information is selected by inhibiting or blocking of information, which surrounds the target stimulus. According to Broadbent, selection occurs early in the process of information processing. The filter selects information on the basis of sensory attributes such as intensity, frequency, or location. According to Norman (1968), initial analysis of information is automatic (not requiring effortful attention). Stimuli activate feature detectors, which then activate units in long-term memory. To "attend" means to activate units in long-term memory that correspond to the attended target. Treisman (1969) suggested that selective attention attenuates signals via a series of filters for all incoming information. Attention is this filtering function, but filtering is sophisticated. Although these filters are limited-capacity processing channels, they can handle complex stimulus attributes. All of these early suggestions concerning models of attention spawned fruitful and productive research efforts that continue to this day. Those that invoke inhibitory processes are more relevant for selectionist memory theory (for reasons described above). I turn now to a brief review of recent empirical findings on inhibitory processes in attentional and memory processing.

Tipper and his colleagues have provided convincing experimental evidence in humans for active inhibitory mechanisms during attentional processing (see reviews in Houghton and Tipper, 1994; Tipper, 1992). These inhibitory processes operate on internal representations of irrelevant or ignored distractors. Tipper reasoned that one should be able to study inhibitory mechanisms by detecting a processing cost associated with use of previously inhibited items. If, for example, a stimulus is inhibited during selection of a target, then using that stimulus in some subsequent processing task should be a little more difficult compared to use of a stimulus that had not undergone inhibition. The processing cost associated with use of previously inhibited items is called the negative priming effect. In standard priming effects, processing of a current stimulus is facilitated by previous exposure to a related stimulus. In negative priming effects, current processing is impeded by exposure to a previously inhibited item. Reaction times are increased to targets that are identical to items that served as distractors—and by hypothesis were inhibited—on a previous trial.

In the standard negative priming experimental procedure conducted by Tipper and colleagues (1992), subjects view a prime display consisting of a red picture (e.g., a kite) superimposed over a green one (e.g., an anchor) and are asked to name what the red picture is. This is the control condition. In a separate "ignored repetition" condition, a red picture (e.g., a kite) is again superimposed over a green picture of a distractor item (a horn) and the subject is asked to name the red picture. "Horn" in this condition is presumed to undergo inhibition as part of naming the target. In the critical "probe" condition, the display is again composed of a red picture superimposed over a green picture and the subject is asked to name the red picture. The red target item in this condition, however, is identical to the distractor in the "ignored repetition" condition (the horn) and the green distractor is a neutral meaningless design. The subject must, in a sense, use the previously inhibited item (from the "ignored repetition" condition) to name the current target item. If the cognitive system uses inhibition to facilitate selection, then reaction times in the critical probe condition should be longer after presentation of the ignored repetition prime than after the control prime. This is exactly what Tipper and colleagues found.

Negative priming effects have been observed in different perceptual modalities and with letters, words, and pictures. Tipper and colleagues (1992) also found that negative priming effects are obtained independent of spatial location of an object and in fact move with the object in question, suggesting that inhibition operates on high-level, object-centered representations. Thus, inhibition apparently works on highly abstract mental representations.

MEMORY AND INHIBITION

Dagenbach and Carr (1994) posited that retrieval itself operated like a selectional/attentional system that enhances the target item by suppressing surrounding distractor items. They had subjects study definitions of extremely low-frequency English words for fifteen seconds each. These newly learned words then served as primes for well-known target words in a subsequent lexical decision task. Prime-target pairs were either semantically related or unrelated. Subjects were instructed to recall the prime's meaning and then to decide whether the target was a real word or not (lexical decision). Accuracy of recall was determined by a post-test recognition test. When subjects recalled the meaning of the prime, lexical decisions on related targets were faster compared to the case when prime and target were unrelated. When subjects failed to recall the meaning of the prime, lexical decisions were slower on related targets as compared to unrelated targets. Lexical decisions to unrelated targets were faster because the non-recalled primes inhibited responses to related targets. Dagenbach and Carr suggest that the demonstrated retrieval inhibition reflects the operation of a center-surround attentional mechanism (analogous to center-surround inhibitory effects in perceptual circuits) that promotes retrieval of weakly activated items by enhancing targeted items (the center) while inhibiting more strongly activated items in the surrounding semantic space (the surround).

Bjork and colleagues (see review in Anderson & Bjork, 1994) have produced further evidence for an active inhibitory-suppression process during memory retrieval. This evidence comes from "directed-forgetting" experimental paradigms. In a typical experiment, subjects learn a list of ten unrelated nouns, after which they are told to forget the list (the forget group) because the wrong words had been given to them. A second group of subjects are given the same list and told to remember the list (the remember group). Both groups then receive a second list of ten nouns, which they are asked to remember. After a brief interval both groups are tested for memory of both lists. Typically subjects in the forget group have difficulty recalling the first list when compared to subjects in the remember group. That is to be expected since they were told to forget that list. Interestingly, forget subjects' performance on the second list is better than remember subjects' recall of the same (second) list. The poor performance of remember subjects on list two is traditionally attributed to proactive interference effects. Apparently instructions (to forget subjects) to forget items on the first list acted to reduce the interference that normally acts proactively on the second list of to-be-remembered items. Proactive interference can only "work" if the items from list one are still activated. Since no proactive interference was observed list one items must have been actively suppressed. Anderson and Bjork (1994) argue that directed forgetting may arise from the action of a suppression mechanism that subjects use to counter the effects of proactive interference. Cognitive inhibition in memory appears to be more than a mere relaxation of activation levels. Rather, an active suppression or de-activation process in memory occurs, at least under conditions of directed forgetting.

FAILURE OF INHIBITORY PROCESSES IN CLINICAL POPULATIONS

Selection's goal is always to extract a signal from surrounding or interfering background noise. There are two major strategies in signal extraction: You can either enhance the signal (e.g., via repetition) while leaving background noise levels unchanged; or you can diminish background noise by inhibiting all aspects of the noisy transmission except those related to the signal. The ideal, of course, would be to enhance the signal while suppressing the noise. Since we humans can function optimally only within rather narrow information processing ranges, we have to regulate the amount of information processing we engage in. We want the signal-to-noise ratio to be optimal. According to information theory, the amount of information carried in the signal can be calculated: $I(\text{s}) = \log2\ 1/p(\text{s})$, where $p(\text{s})$ is the probability that a particular signal will occur and $I(\text{s})$ is the amount of information (in bits) transmitted by that signal. The probability that a signal will occur, of course, depends on noise level or the reduction of uncertainty associated with the signal. If we are unable to keep signal and noise levels manageable, we suffer. The inability to screen out irrelevant noise or stimuli can lead to mental breakdown. Selective attention allows us to control the amount and quality of information we are exposed to.

Inhibitory mechanisms in attention and memory support this selection process and are ecologically real. If selectionist processing depends on inhibitory mechanisms, impairment in these mechanisms should result in selection and attentional failures and/or increased vulnerability to cognitive interference. The patient with such an impairment would be unable to screen out irrelevant stimuli and normal cognitive processing would be seriously impaired. Interestingly, reduced (relative to healthy controls) negative priming effects (indicating inefficient inhibitory processes) have been observed in clinical populations who evidence increased vulnerability to interference during cognitive tasks. These populations include children with attention deficit disorder, depressed patients, schizophrenics, patients with Alzheimer's disease, and the elderly. While I could find no study of negative priming in patients with amnesia due to Korsakoff syndrome, it is likely that they will exhibit reduced negative priming effects as well. These experimental and clinical findings with negative priming effects suggest that inhibitory mechanisms assist in selection of a target stimulus by reducing the interference from competing, irrelevant distractors. They also point to the central role inhibition and selection play in the mental life of an individual. When dis-inhibition results in interference effects in the memory system (as in Alzheimer's), personality itself gradually disintegrates. When the locus of the interference is in attentional systems (as in attention deficit disorder), disorganization and learning deficits are common. Interference effects appear to globally influence both cognitive and emotional processing in patients with brain-based psychiatric syndromes, such as schizophrenia or autism. Patients with frontal dysfunction evidence greater interference scores on tests that tap inhibitory processes such as the Stroop color-word test, and the Wisconsin Card Sort test. Most of the clinical populations who show reductions in negative priming effects also perform poorly on these frontal tests. Thus, these experimental and clinical paradigms appear to be measuring a real and potent inhibitory process, which performs essential functions for the cognitive system.

INHIBITION OUTSIDE THE LABORATORY

Vulnerability to interference in everyday cognitive processing can be a serious problem. The work of Loftus and colleagues (see review in Loftus, Feldman, & Dashiell, 1995) on what are now called false memories show that even minor interference effects can profoundly influence retrieval from episodic memory systems. In a typical experiment, these researchers have subjects watch a film of a car crash. Later all subjects are asked the same question about the film except, for some, the word "smashed" was changed to "bumped," and so on. When asked a week later whether there had been any broken glass in the film, subjects who had been given the word "smashed" reported incorrectly that there had indeed been broken glass in the film. These results suggest that everyday instances of remembering are colored by current salient stimuli, which potentially interfere with accurate recall.

UTILIZATION OF ASSOCIATES TO CONTEXTUALIZE THE CUE

Bergson's model of remembering not only holds that distractors are generated and then, if irrelevant, suppressed, but also predicts that these distractors are utilized in the selection process. Indirect evidence for activation then inhibition of context-appropriate associates to a target comes from lexical-ambiguity resolution studies. In a typical example (Swinney & Prather, 1989), a polysemous word is presented in a sentential context that primes one of the meanings of the word ("The dog gave a loud bark and growled"). Subsequently, reaction times are probed for close associates of bark (dog) and bark (tree). Analysis of these reaction times suggests that, at least initially, multiple senses of an ambiguous word are generated and then later the meaning most contextually appropriate is selected. Thus, even if the sentential context strongly requires a single meaning, it appears that all possible meanings of the word are generated before selection can proceed. This result, of course, is predicted by a selectionist account and it implies that the generation of associates to a target is obligatory and functions to narrow the search/selection procedure. Findings from these lexical-ambiguity resolution studies are robust, and similar phenomena involving selection via inhibition and suppression of alternatives have been demonstrated in virtually all other aspects of language comprehension (see review in Gernsbacher & Faust, 1995).

Houghton and Tipper (1994) point out that there is abundant evidence for active evaluation of apparently non-attended or suppressed stimuli in our everyday lives. This active evaluation of to-be-inhibited information presumably serves to focus attention on the target. When, for example, we notice the offset of a background noise to which we had become habituated, it indicates that we must have been processing and actively monitoring that background information. Monitoring background noise levels helps us to adjust attentional levels to the task at hand. Similarly, take the case of unconscious plagiarism. Here the writer uses material someone else authored and takes it as his own. Since the plagiarist has no memory of seeing the "borrowed" material, we have to assume that it had been only casually processed. Yet the plagiarism proves that the apparently ignored information was actively evaluated.

MATCHING AND SELECTION

What about the selection process itself? As mentioned above, attentional processes must surely guide the selection mechanism itself. Bergson and most other selectionist authors conceive the selection process as a kind of selective attending and matching algorithm. Items whose properties fail to match the target item's properties trigger a mismatch signal that then shifts the balance of excitatory/inhibitory signals in the attentional system toward inhibition. This inhibitory shift, when repeated with each distractor, results in a wave of spreading inhibition, which should be detectable with experimental techniques. Näätänen (1982), after a thorough review of the literature on the event-related potential component known as processing negativity, concluded that the component was produced by a match/mismatch process, which compares sensory input to what he called the "attentional trace"—what we have been calling the target item. Once distractors are actively suppressed, excitatory mechanisms boost the activation levels of the target item, and it thereby enters into the subject's online, aware processing systems.

This cursory review of the experimental evidence for inhibition effects in memory suggests that each of the six descriptive elements (see p. 46) which make up Bergson's selectionist memory theory, are supported by the evidence. Inhibition is real, it operates on high-level mental representations, and it is used to facilitate selection of a target item. Since inhibition would not be required if no distractors existed, inhibition implies that the mind generates copies, associates, and distractors of a target image during selection of a target item. Part of the reason for generation of these distractors is that the distractors facilitate selection of the target by contextualizing the cue or narrowing down the search. Once selected, target images are entered into long-term memory stores. All of these processing steps, furthermore, are supported by relatively well-described neuropsychological systems.

NEUROPSYCHOLOGICAL MODELS CONSISTENT WITH BERGSONIAN REMEMBERING

Pribram and McGuinness (1975) and Tucker and Williamson (1984) review the neuropsychological evidence for a processing system much like that proposed by Bergson. Their review of the literature on the neuropsychology of attentional processing suggests that an arousal system, supported by activity in norepinephrine networks (interacting possibly with serotonin) and centered in posterior parietal and right-hemisphere neural systems, is counterposed to an activation network supported by ascending nigrostriatal and mesocortical dopaminergic systems (interacting possibly with acetylcholine) and centered in anterior and left-hemisphere neural networks. The two systems are in mutual inhibitory balance and work together during selective processing operations.

The posterior noradrenergic system responds preferentially to novel stimuli and promotes habituation to repetitive stimuli. In information processing terms, this posterior system acts phasically to reduce informational redundancy in working memory stores. With a bias toward habituation and minimal redundancy, representational formats are expansive and externally controlled. The preference for novelty implies an external and data-driven process, which fills working memory stores with a vast array of new items. The breadth of informational sampling in this case implies little or no transfer of information into long-term memory. Sampling a broad array of unique items also implies a holistic or global configurational analysis of the items presented to working memory. The reliance on external information sources places the control of ongoing behavior in the external environment and enhances exploratory behavior.

The anterior dopaminergic system is tonically active in the service of motor readiness. Dopaminergic activity facilitates selection of specific motor acts. An increase in levels of dopamine produces a corresponding increase in the restrictions on the range of motor acts produced. With high levels of dopamine, only high-probability motor acts are produced in repetitive fashion. Eventually, only stereotypies are emitted. In information processing terms, the anterior activation system acts to promote redundancy in working memory stores. A redundancy bias restricts change or flow in information networks. Representational scope is restricted and narrow, with less information in working memory stores. Information from memory stores is, therefore, available to guide ongoing behavior. Control of behavior is shifted from an external locus to an internal locus. Internal control allows the organism to resist external distracters or interference. When motivation levels are high, the bias toward increasing redundancy leads to an increased likelihood of repeating past behaviors; thus, this supports habit formation. The bias toward redundancy also promotes an increased degree of processing on each event held in working memory. Depth of processing on a few elements is counterposed to the arousal system's superficial processing on a large array of elements. A succession of focalized processing cycles on a series of redundant elements produces greater differentiation of the series of elements.

The anterior and posterior attentional systems constitute a processing cycle that is very much like Bergson's selectionist remembering cycle: The arousal system, with its preference for novelty and its ability to habituate to non-novel stimuli, allows for proliferation of a vast array of novel elements within the cognitive system. Initial global analyses are conducted but no elements enter into long-term memory; that is no elements are selected. After the proliferation phase, the activation system with its ability to maximize redundancy selects out (via analytic and depth-of-processing abilities) redundant elements/attributes that are then transferred into long-term memory. These redundant elements may be thought of as James's substantive portions of the stream of thought (see Chapter 5). As Tucker and Williamson put it, "By decrementing the response of perceptual networks to repetitive input, a habituation bias enhances the brain's responses to novelty. In an ongoing stream of perceptual processing, this effect decreases the attention allocated to a given event and maximizes the number of unique events that receive attention. In contrast to a redundancy bias, which causes a few events to be repetitively encoded in short term memory, a habituation bias saturates memory stores with a broad range of unique events" (1984: 203). Thus, we come equipped with a set of neuropsychological systems whose operations are compatible with the basic descriptive elements in Bergson's selectionist memory theory.

INHIBITION AND SELECTION IN NEURAL NETWORK SIMULATIONS

Neural network modeling has proven itself useful in explorations of basic cognitive functions. Inhibition plays a central role in these connectionist models and their closely allied distributed memory models. Computation typically takes place in a network of excitatory and inhibitory connections among simple units. Each unit or node has an activation level that is a function of excitatory and inhibitory inputs it receives as well as its own recent activation history. The units are arranged in layers, most commonly in an abstraction hierarchy with units representing simple feature detectors lower in the hierarchy and units representing complex schemas or scripts higher in the hierarchy. In an autobiographical memory network, one might find event units in the lowest layer, episode units in the next layer, schemas in the next, and stories in the topmost layer. The network computes by activating an outside layer (thus capturing either bottom-up or top-down processing effects), and activation spreads throughout the network. Connections within a layer are inhibitory (on the model of lateral inhibition in the CNS) and those between layers are excitatory. Since connections between layers run in both directions, processing is interactive. Trial and error learning or retention may result from procedures (e.g., the delta rule, back propagation), which change connection weights between units in response to feedback from some criterion target activation levels. Although this capacity to generalize is one of the most desirable features of a connectionist model, it nevertheless leads to rapid forgetting or what has been called "catastrophic interference" effects in the network. Attempts to solve the interference problem include careful attempts to enhance inhibitory powers of the network. Neural network connectionist modeling, in any case, seems broadly compatible with selectionist approaches to cognition and may yield powerful tests of selectionist approaches to Mind and memory.

MODERN SELECTIONIST RESEARCH PROGRAMS

There have been a handful of accounts of basic cognitive and memory functions that explicitly draw on selectionist principles—though there is no complete treatment of cognitive memory theory. According to these recent accounts, there are two major types of memories: those that become available during the generation process (internally generated forms) and those that are linked with the amplification process. Under these two broad subdivisions lies a whole range of memory types, which have already been referred to. The internally generated memory forms emerge as "prerepresentations" during brain development and probably coincide with what is usually called innate structures/processes or mental representations. Thus the organism comes equipped with a plenitude of representations that are experienced as "memories." In the adult Changeux and Dehaene (1989) suggest that innate "prerepresentations" are probably promoted by the spontaneous activity of neurons known as cellular oscillators. These neurons require no external source of stimulation for their activity. Presumably, innate memories include everything from instinctual preferences to specialized pools of knowledge that support various cognitive procedures to forms corresponding to Jung's archetypes. At a minimum, prerepresentations would have to correspond (or refer or be tuned to) to something like Gibsonian invariants or affordances in the outside world.

The selection process itself has been conceived in a variety of ways by modern investigators. In general, the idea is that the item chosen/selected (say, a Hebbean-type assembly of neurons) is stabilized while all other items/possibilities are allowed to relax back into formlessness or low levels of activation, or are actively inhibited. Edelman (1987) proposes re-entrant signaling and recategorization as basic for mnemonic phenomena. Several theorists endorse the notion of resonance as the principle underlying activation of stored representations—what we are calling here "selection." A memory is stabilized if it resonates (roughly: matches) with an item from either the environment or from another part of the cognitive system. Young (1979) suggests that reinforcement is involved in the stabilization process once a selection is made. Piatelli-Palmarini (1989) suggests that selection is essentially a triggering of a set of switches or parameters, which then determine the set of possible computational procedures available to the system.

Edelman and colleagues (see Edelman, 1987; Sporns & Tononi, 1994; Reeke & Sporns, 1990), in perhaps the most developed modern research program that explicitly adopts the selectionist paradigm, have proposed an account of brain and memory function that they call the Theory of Neuronal Group Selection. These investigators point out that the most salient fact of nervous system organization is its variability. Brain functions are carried out in parallel by units that are slow, numerous, and densely interconnected. The many different kinds of processing elements, as well as the myriad ways in which the elements are interconnected, confer an extraordinary degree of flexibility and variability on the functional elements of the nervous system. This flexibility is related to Edelman's use of the mathematical concept of "degeneracy" to refer to the ability of various neuronal populations to accomplish similar functional goals. Nervous systems have evolved in such a way that variance is an essential part of their adaptation.

Diverse neuronal elements and circuitry are generated during embryonic development. Selection operates to preserve and enhance the activity of parts of the developing nervous system that have a competitive advantage over other parts of the nervous system. Cells begin to form groups or networks that tend to share similar functional properties. Neuronal networks that survive the neuronal elimination process are called the primary repertoire and subserve activities that are in some way useful to the individual organism.

Once the primary repertoire is formed, selectional processes begin to shape and refine neuronal functioning. Selection acts on neuronal groups to modify the strengths of individual connections both within groups and between groups and networks. Within-group neuronal responses are fine tuned to local environmental contingencies and between group responses compete for access to the motor efferents of the brain, thus influencing the output of, or frequency of, different types of behaviors. This secondary shaping and tuning process results in the formation of a new group of interconnected neuronal networks, called the secondary repertoire. Despite these developmental selectional events, variability in neuronal repertoires is never entirely depleted. Mechanisms (e.g., proliferation of new synaptic connectivity patterns) exist to continually introduce variability into the system.

Re-entrant signaling is the hypothesized neuronal basis for higher-order categorization processes. It refers to the exchange of signals between neuronal groups, which promotes similar response profiles for each of the participant groups. The groups establish similar response profiles by exchanging signaling patterns. When an exchanged signal returns to its original source-group, it tends to force response patterns in each part of the network to form a single unitary whole. This type of recursive re-entrant signaling may contribute to the formation of a consistent set of responses across time, resulting in recategorization and memories.

Given these facts concerning the enormous variability that exists in structure and processing elements of the central nervous system, models of Mind that gloss over this variability are bound to fail. According to Reeke and Sporns (1990), functionalist accounts of mental function have underestimated the problem of variability. Functionalism holds that mental processes are best understood in terms of functional processes like symbol manipulation or computations on symbolic tokens. The brain, in effect, is best understood as a Turing machine—a device bearing encoded information and a fixed set of rules to manipulate that information. Mental states are determined by the causal properties associated with computations on primitive symbolic tokens. Functionally identical computations give rise to identical cognitive states. Computational/symbolic processes operate by carrying out sequences of elementary steps on strings of primitive symbolic tokens in order to produce a relatively invariant mental result. If these processes are to work, it is therefore essential that the precise sequence of these steps be reliably maintained throughout the life span of the individual organism. But Reeke and Sporns argue that this "maintenance requirement" for the functionalist model of Mind and memory is impossible to satisfy. The very substance of the Mind/brain is variability. Adaptation would be impossible without this variability. Nor can invariant functional sequences, required by functionalist theory, be transmitted across generations via the genetic code—at least no one has suggested any mechanism that would allow such a transfer. There is, furthermore, no known mechanism to communicate desired output values (or goals) of the organism to individual cells or synapses in the brain. Models of brain function that depend on specific learning algorithms to accomplish such feats are therefore inappropriate or inaccurate models of brain function. While selectionist models of brain function have none of the problems which functionalist models contend with, there is still the problem of showing that selectionist mechanisms can produce something like cognitive processing. Reeke and Sporns (1990) have described their efforts to show that the principles of Edelman's selectionist theory can be simulated successfully in a neuronal network computer model or recognition automaton. They call this machine Darwin III. Darwin III has, in fact, been shown to be able to perceive, categorize, and respond to a small set of objects in its immediate environment. While these accomplishments are no better than those associated with neural network models/machines, the point here is that they were achieved via selectionist principles.

Calvin (1996) has recently produced a detailed theory on how Darwinian selectionist principles might operate in the neocortex. According to Calvin, there are six essentials to any true Darwinian process that can produce something like intelligence. First, there is a characteristic pattern or code (e.g., genes), which acts as the unit of selection. In neocortical processing, Calvin suggests that the requisite pattern is a spatiotemporal firing pattern within a 0.5 mm hexagonal slice of cortex. He claims that the firing pattern involves pyramidal neurons in superficial layers II-III (usually believed to be the cortical site of associative memory networks). Second, the pattern must be copied. Point to annulus connectivity patterns between large pyramidal neurons allow them to recurrently excite one another such that they entrain a third and fourth to initiate a "triangular array." The recurrent excitation functions to clone all active arrays within a hexagonal unit (hexagonal cloning). Third, the copying process must occasionally produce errors or variants of the original code. Here I find Calvin's suggestions unclear (he claims there are noncloning hexagons that act as barriers or error-correction devices to the copying process, but relaxing activation levels of these "gateways" is just as likely to produce more copies rather than errors). I assume errors can be introduced just by allowing background noise levels to increase. Fourth, the pattern/code and its variants must compete with one another over access to a limited workspace. For Calvin, the limited workspace appears to be the cortex itself. Fifth, the competition is biased by environmental factors giving one pattern more of the workspace than another. Long-term memory serves as part of that environment. Its network of associations creates basins of attraction that favor one pattern over another. Sensory inputs also play a role. Sixth, and finally, there is a skewed survival to reproductive maturity such that new variants are generated around the more successful of the current patterns. This is Darwin's inheritance principle. Here Calvin seems to suggest that those patterns that win out in the competition occupy a larger area of cortex. Given the larger area there is a greater likelihood that new variants will be produced from successful old variants simply because the old variants occupy more space than unsuccessful variants. The greater the space, the greater the chance that errors will be produced from that space.

It is impossible to do justice to Calvin's ingenious proposals here. Like Edelman's ideas, these proposals are plausible and thus testable using current techniques. The program of research loosely referred to as Mental Darwinism now has some detailed proposals on the table concerning the evolution and functioning of human intelligence. This set of proposals boil down to the handful of selectionist principles described above. They are the same principles that hold true throughout the biological world. If we can demonstrate that Darwin's principles apply in the human neocortex, we will have brought the laws of science, as Darwin once said, "into the citadel itself."

In particular, the modern experimental evidence supports Bergson's model of selectionist remembering. This is quite an achievement for a philosopher working at the turn of the last century. Bergson, however, did not have much to say about the relationship between memory and identity. If recollection increases the rememberer's range of useful and free action, as Bergson argues, doesn't this mean that remembering is in service to a particular, personal consciousness? If remembering is always colored or biased for a particular rememberer, doesn't that mean that veridical remembering is compromised or excluded? How does memory give rise to or impact on personal identity? Selectionist approaches require cumulative retention of selected images. This cumulative effect gives rise to the historical feel of memory. Memory develops or evolves, just as people do. Thus, problems of memory and personal identity are inextricably intertwined, but Bergson had little to say about identity. James, on the other hand, had a lot to say about self and consciousness. Before we turn to James's works on memory and self, we need to complete our review of the role of inhibition and selection in cognitive and memory mechanisms. This review will necessarily involve a discussion of the role of the frontal lobes in memory and inhibitory processes.

CHAPTER 5

Frontal Lobes, Memory, and Inhibition

SELECTION AND INHIBITION

Bergson argued that memories are never lost. We are, in fact, always steeped in an infinity of memory images. Inhibition keeps us from noticing most of them or being mentally crushed by them. Selective inhibition allows us to focus on a tiny subset of these memory images during the selection process. Inhibition also serves to preserve the selected subset of memories by protecting them from interference by other memories, which seek to displace them. Relaxation of inhibition has the effect of flooding the cognitive system with an array of memory images and their associates. Inhibition plays a central role in Bergsonian memory theory, and it is crucial for cognitive selectionist theory in general. It constitutes indirect but convincing evidence for the existence of "internal distractors," which participate in cognitive processing, including memory. Inhibition and interference effects in cognitive processing would not be seen if selection from an array of distractors were not taking place. Distractors would not be present if they had not been generated or allowed by Mind. "Selection" is always from an array of alternatives, some of which are inevitably irrelevant. Inhibitory mechanisms are, in fact, ubiquitous in brain and cognitive processing (Dagenbach & Carr, 1994; Dempster & Brainerd, 1995). In this chapter, I discuss the evolutionary and neuropsychological correlates of cognitive and behavioral inhibition mechanisms, and then relate this material to selectionist memory theory.

EVOLUTION OF INHIBITION

Cognition

Inhibition makes possible the so-called "executive functions" of the human brain. Executive functions include short-term working memory, manipulation of contents in short-term and working memory, internal speech, development of anticipatory sets, deferred gratification or withholding of prepotent responses, goal-directed, purposive behaviors, planning, and other similar behaviors. Clearly, all of these executive functions assist in personal self-regulation of behavior. Personal autonomy is dependent on their proper functioning and they, in turn, are dependent on inhibitory mechanisms. Goal-directed, purposive behaviors are not possible for an individual who is incapable of resisting interference from other competing goals. Retaining information in the short-term memory workspace is possible only if that information is not immediately displaced by competing, interfering stimuli. Inhibition allows us to ignore irrelevant stimuli and to attend to relevant stimuli. It is crucial for concentration and deliberative thought. No sustained attention to a single train of thought is possible if one's thought is constantly falling prey to distractors. Concentration requires the ability to ignore or screen out distracting and irrelevant information. The bare ability to hold some thing in one's mind requires the ability to resist displacement of that mental object by some other mental object. Resistance to interference is a prerequisite for focused, concentrated mental processing.

The link between cognitive inhibition and the ability to concentrate is important for a theory of the evolution of Mind. Take, for example, the fashioning of a tool kit for the production of implements and weapons. Many scholars have pointed to tool making—especially when it became innovative and inventive—as an important correlate of the evolution of the Mind. It seems highly unlikely that a set of tasks requiring sequencing, planning, fine motor control, problem solving, reasoning, and modeling could be performed by a proto-human hominid with only poor abilities to resist the effects of cognitive interference. Innovative tool-making requires a kind of exploratory behavior in the cognitive realm. Exploratory behavior, in turn, requires the ability to inhibit established, overlearned responses so that the animal can orient toward novelty.

Cognitive inhibition may have also been important for the evolution of language abilities. The acquisition of language in the child requires suppression of babbling and other emotional vocalizations; a similar process may have occurred in the evolution of language. Although there is some overlap in the anatomical circuits supporting vocalizations in monkeys and in humans, there appears to be greater cortical control of vocalizations in humans. Broca's area, the motor and premotor areas, and the supplementary motor area are especially important for human vocalizations, while the cingulate gyrus, limbic sites, and the central and periaqueductal gray areas in subcortical and brainstem sites all contribute to control of vocalizations in both monkeys and humans. The systematic inhibition of emotional vocalizations via neocortical controls may have been one consequence of a more generalized shift toward cortical inhibition of limbic and emotional responses in hominid evolutionary development.

Speech production, in any case, requires extremely sophisticated inhibitory machinery to function properly. Speaking a sentence or message is the end result of a fairly large number of cognitive selections, such as which lexical items to use, which syntactic structure to use, how much information to convey, how much information to conceal, intonational and phrasing decisions, where to place pauses, prosody and stress, and a host of other selections. Traditionally, linguistic procedural decisions have been divided up into two broad classes. Paradigmatic decisions involve selection of a target item (word, phoneme, phrasal structure) from among an array of possibilities that will convey the intended message. Syntagmatic decisions involve selection of the correct order of elements in a given linguistic sequence or the correct placement of a target unit into a target sequence. In both types of decisions, the correct target unit must be selected over an array of competitors and selection must occur rapidly, on line, as the subject is speaking. It is difficult to see how this kind of cognitive processing could be accomplished without dual mechanisms of inhibitory and excitatory effects. Indeed, most neural-network models of speech production employ a combination of activation and inhibitory circuits to adequately model ongoing speech production.

Bronowski (1977; see also Barkley, 1997) argued that the evolution of the ability to inhibit and delay responding resulted in four cognitive abilities that underpin human language. These are prolongation, separation of affect, internalization, and reconstitution. Prolongation is the ability to refer cognitively to past and future events in time. The ability to delay gratification in order to obtain a future reward apparently contributes to the time sense in human beings. The prolongation of reference across time epochs requires a special form of memory that situates the self in time. Wheeler, Stuss and Tulving (1997) call this special form of episodic memory "autonoietic consciousness" and link it to frontal lobe functions. I will return to this issue below. The separation of affect refers to the partial segregation of emotional charge or valence from the content of a vocalization or message. Internalization refers to the ability to internalize speech and language. As just mentioned, inhibition allows for suppression of the external manifestations of emotional vocalization. The imposition of a delay between an event and a person's response to it allows time for the event to recruit several brain areas into its evaluation process. Cross talk between these brain areas can then proceed. Internalization of speech appears to be important in the child's development of self-regulatory skills. Reconstitution involves the breakdown of events into more basic units for analysis, and the synthesis of these units into new messages. Analysis presumably involves decomposition of events into the standard thematic roles of who did what to whom. The reconstitution of units into larger messages eventually yields the hierarchical structure of language as well as its compositional attributes. Bronowski (1977) credited the frontal lobes with the abilities to reconstitute, to internalize speech, to prolong reference across time epochs, and to separate affect from the content of messages. We will see that the frontal lobes are also implicated in social and emotional inhibitory effects.

SOCIAL AND EMOTIONAL BEHAVIORS

Freud argued that the ability to inhibit sexual and aggressive impulses is a prerequisite for social and civilized behavior. If individuals can derive real benefits (e.g., a larger "return" later) by learning to inhibit current appetitive or consumatory responses, then natural selection would favor those individuals with the ability to delay gratification of impulses. What kinds of evolutionary benefits would strong inhibitory abilities confer on an individual?

One benefit would be an enhanced ability to engage in deliberative and reflective thought. These individuals would have better planning and analytical skills. Social cooperation would be easier in a group of individuals who could moderate their sexual, aggressive, and appetitive responses.

Another benefit would be an enhanced ability to socially compete with, deceive, and manipulate conspecifics (Bjorklund & Harnishfeger, 1995).

The most spectacular example of deception among hominids is concealed ovulation. Unlike most other mammals, human females conceal the time of their ovulation, thus providing no external sign of their impregnability. Most scholars who have studied the evolution of concealed ovulation suggest that the concealment would have the effect of eliciting a greater investment of time and resources from a male since the male would never know for sure when the female is receptive sexually. Concealed ovulation may imply a gradual transfer of control of sexual responding from limbic sites to cortical sites. It surely implies an ability on the part of the female to utilize potent physiological and psychological inhibitory powers in order to manipulate the male. Whether or not concealed ovulation evolved as part of a strategy to deceive and manipulate males, it does seem to have that effect.

Inhibition of social and emotional behaviors in service to deception begins quite early in humans. Parent-offspring conflict theory (Trivers, 1985) begins with the idea that the genetic interests of parents and their children do not entirely overlap. Each child is interested in eliciting a maximum amount of resources from the parents, while the parents are interested in dividing their time and energies roughly equally among the children. If, on the other hand, one child appears to be a particularly healthy genetic "bet" the parents may decide to invest more heavily in that child at the expense of others. Children, therefore, devise deceptive strategies in order to manipulate the parents into investing more resources in each one of them. Children must utilize their inhibitory powers to behave in ways that will elicit greater support from the parents. Children may advertise their neediness by appearing more immature than they are. If parents do not indulge these needy and weepy children, they may switch their strategy and begin to advertise their independence by acting more mature than they are, and so on. At the extreme, the child may act seductively toward an opposite sex parent, thus confirming, in a sense, Freud's claims about Oedipal and Electra complexes. The choice of what form the deception will take depends on the reactions each child gets from the parents. But the deception, of course, requires sophisticated inhibitory abilities to work.

Byrne and Whiten (1988) have argued that an important driving force in the evolution of human intelligence has been social competition between individuals. They term the ability to engage in such competitive dynamics "Machiavellian intelligence" to emphasize its political characteristics. Machiavellian social maneuvers include the formation of coalitions, manipulation of conspecifics, deceptions of others, the ability to self-deceive, cheating, and other political machinations. All of these social operations require extremely sophisticated inhibitory processes. Manipulation and deception of others requires the ability to suppress or delay sexual, emotional, and aggressive impulses, or at least to subliminate them. Sublimation, in turn, requires inhibition of immediate gratification of those impulses. Trivers (1985) has pointed out that the best deceivers and manipulators of others are often the self-deceived. It is difficult to lie convincingly when one knows that it is a lie. Human beings are very adept at catching liars. Subtle signs, such as a facial expressions, sweaty palms, or a tremor in the voice, may give away a liar. But if the liar believes the lie himself, none of these subtle clues to the deception would be present and the lie would go undetected. To self-deceive, one would need to suppress information that contradicts the lie. One would have to sequester the truth away from awareness. One would have to edit one's own autobiographical memories in order to self-deceive. Only certain autobiographical memories would be selected as real autobiographical memories. Only a select few would be eligible to enter into awareness or consciousness. Deception of others and of self might therefore have been one of the evolutionary sources of the exceptional inhibitory powers associated with modern humans.

HOMINID EVOLUTION

At what point in the evolution of modern humans did these exceptional inhibitory powers manifest themselves? What we know of hominid evolution suggests that the process must have been gradual, though punctuated by a great leap forward about 50,000 years ago. The first step in the evolution of modern *Homo sapiens* from the apes is generally considered to be development of upright posture. Though Leakey's discovery of Australopithecine footprints dating from about 3.5 million years ago places a lower limit on the emergence of bipedalism in the hominid line, the ability probably evolved some 6 million years ago. Presumably, reliance on the sense of smell needed to be suppressed or inhibited for bipedalism to work. By about 2 million years ago, there were several species of Australopithecines including *Homo habilis*. Its cranial capacity was about half that of modern humans and it fashioned some relatively crude stone tools. Nevertheless, the fossil evidence suggests the presence of a prominent Broca's area and inferior parietal lobule in the left hemisphere of the Habilene brain suggesting some language-related capacity in this early hominid species. About 1.5 million years ago, *Homo erectus* appeared with a slightly larger cranial capacity that that of *habilis* but with substantially more advanced tool-making abilities. Anatomically modern humans emerged between 150,000 and 300,000 years ago, but we still do not see comparatively modern tool-making abilities and complex social organizations until some 50,000 years ago. All of the innovations we associate with distinctively human abilities, such as sophisticated tool making, traditions and styles in artistic production, trade between communities, and other signs of complex social organization did not appear until some 50,000 years ago. Lieberman (1984) argued that only modern *Homo sapiens* developed a vocal tract that could produce the rapid speech sounds we know as language today. Even though early hominids apparently possessed the brain prerequisites for language, these linguistic abilities could not translate into modern, fluent speech until the vocal tract was shaped into its final modern form; this occurred only with anatomically modern humans. This modern ability to rapidly and automatically produce the full range of speech-related phonemes may have constituted a decisive competitive advantage over our relatives the Neanderthals. Although *Homo sapiens* had a smaller cranial capacity and a slighter build than that of their contemporaries, the Neanderthals, Neanderthals became extinct by about 35,000 years ago.

There is probably no way to know when sophisticated inhibitory abilities took hold in the hominid line. We know, however, that chimpanzees and other close relatives engage in relatively sophisticated social deception and manipulation, so inhibitory capacity must be an extremely ancient capacity. Probably the acquisition of advanced and powerful forms of cognitive inhibition occurred gradually with each hominid species contributing a new and more complicated addition to the inhibitory repertoire. Yet hominids have taken deceptive practices far beyond the games that our nearest relatives, the chimpanzees, play. And the neural supports for inhibitory powers have also evolved significantly with each new hominid species. At some point, these inhibitory abilities became powerful enough to make complicated cognitive processing possible by protecting against even high-level mental interference effects such as irrelevant memories, images, fantasies, and emotions.

In summary, it is no exaggeration to claim that the increase and extension of inhibitory powers of the human brain have been crucial in the evolution of the human Mind. Most accounts of inhibitory mental mechanisms refer to the frontal lobes as the neuropsychological site that mediates inhibition and it is to these that we now turn.

NEUROLOGY OF INHIBITION

The evolution of inhibition in human beings is largely dependent on the evolution of the frontal lobes. Fuster (1989) proposed that the prefrontal lobes participate in three fundamental functions: (1) working memory which serves to keeps relevant or salient information on line until it is no longer useful, (2) maintenance of a preparatory set, and (3) inhibitory control over distraction. Goldman-Rakic (1994) argues that the prefrontal lobes access, maintain, and update memories of relevant stimuli in "working memory" in order to guide ongoing behavior. By relying on memories to guide behavior, individuals are not influenced by external stimuli only. Several lines of evidence suggest a central role for the frontal lobes in inhibition, including cognitive inhibition.

Anatomy

The frontal lobes increase in size and connectivity with both phylogenetic and ontogenetic development (Fuster, 1989). They receive projections from the mediodorsal nucleus of the thalamus and make up the large expanse of cortex in the anterior portions of the brain. They comprise primary motor cortex, as well as premotor, supplementary motor, and prefrontal areas. All of these areas send inhibitory efferents onto their sites of termination. The motor-premotor areas comprise Brodmann areas 4, 6, parts of Area 44 (Broca's area), and the frontal eye fields. The prefrontal areas are further subdivided into two large functional regions. The first prefrontal region includes paralimbic cortex of the anterior cingulate region and orbitofrontal cortex. The second prefrontal region is generally referred to as the dorsolateral prefrontal region and includes Brodmann areas 9, 10, 11, 12, 45, 46, and 47. Damage to the primary motor area is associated with paralysis. Damage to the premotor area is associated with disturbances in the organization of movements. Left-sided premotor lesions in Broca's area usually lead to a nonfluent aphasia with preserved comprehension of speech. Supplementary motor lesions are associated with difficulties in the initiation of movement.

When the difficulty involves the initiation of speech, a transcortical motor aphasia may be present. Prefrontal lesions are also associated with a host of processing deficits that are usually characterized by the inability to resist the effects of interference.

Frontal Connectivity Patterns

Much of the functioning of the frontal lobes can be inferred from their patterns of connectivity. The dorsolateral division exhibits greater interconnectivity with basal ganglia sites than does the orbitofrontal division; the orbitofrontal division is densely interconnected with limbic sites. Prefrontal sites are interconnected with virtually all other cortical sites. Their projections to other cortical sites are typically onto layers I, V, and VI—the output layers of the cortex. This implies that the frontal lobes are in a position to inhibit and regulate the output of other cortical regions.

Cummings (1993) identified five patterns of subcortical connectivity: a motor circuit originating in the supplementary motor area, an oculomotor circuit with origins in the frontal eye fields, a dorsolateral, a lateral orbital and anterior cingulate circuits. Each circuit projects to striatal structures, then to globus pallidus and substantia nigra (the dopaminergic nucleus), then from these two structures to specific thalamic nuclei and a final link back to frontal lobes. Three distinct neurobehavioral syndromes are associated with lesions to each of the three prefrontal-subcortical circuits. The dorsolateral prefrontal syndrome is characterized primarily by executive function deficits in planning, working memory and self-regulation. The orbiotfrontal syndrome is characterized by dramatic personality changes. Many of these behavioral changes involve disinhibition of emotional and prepotent responses. A prepotent response is one that has been previously associated with a strong reward. Medial (anterior cingulate) lesions are associated with motor apathy (including akinetic mutism with bilateral lesions) and failure of response inhibition on go, no-go tasks. The prefrontal cortex is the only cortical region which directly regulates neurotransmitter nuclei in the brain stem as well as cholinergic nuclei in the basal nucleus of Meynert. Because these ascending neurotransmitter projections project diffusely to all regions of the cortex, the frontal lobes can influence global cortical arousal levels.

Sanides (1969) and Pandya and Barnes (1987) noted that progressively changing cellular fields represent waves of finer differentiation of cellular layers during evolution of the brain. The architectonic trends originate from separate subcortical moieties. A medial archicortical trend originating in the hippocampus and the proisocortex of the medial wall of the frontal lobe gives rise to the anterior cingulate cortex and to supplementary motor cortex and inferior frontal sulcus. The lateral trend begins in the olfactory cortex, peripaleocortical proisocortex of the insula, and proceeds superiorly up the lateral convexity, the superior temporal lobe and inferior parietal sites and thence to somatosensory area II. Again, the trend in each case is for greater differentiation of cellular layers as the trend proceeds toward its "termination" sites. The two trends interact in dosolateral prefrontal cortex. Goldberg (1987) argues that the lateral trend is overepresented in posterior portions of the brain, especially the parietal cortex, and thus the parietal and frontal lobes operate in a kind of mutual inhibitory balance. The great superior longitudinal fasiculus connects frontal and parietal sites, allowing for nuanced interactions between anterior frontal systems and posterior parietal systems. The frontal lobes rely on processing of internal information, including memories, in order to guide behavior, while the parietal lobes specialize in processing of externally based sensory information, in order to guide behavior. Both sources of information are needed to self-regulate one's own behaviors but internal mnemonic sources are crucial to resist enslavement to external and salient events. We will return to this issue below.

Neuropsychological Evidence

One of the most consistent deficits associated with prefrontal lesions is disinhibition. In humans, patients with prefrontal lesions are especially vulnerable to interfering stimuli and they fail to inhibit prepotent responses. The ability to concentrate and to voluntarily deploy attentional resources is impaired after prefrontal lesions. Typical tests that tap frontal abilities to resist effects of interference are the Wisconsin Card Sort Test (WCST), the Stroop color-word test, and the delayed-response task. In delayed-response tasks, the subject is presented with two adjacent cups. The subject observes the experimenter place a reward under one cup and then pull a door down to hide the cups. After a delay (the length of the delay is varied) the door is raised and the subject is required to retrieve the reward. Subjects with frontal lesions (especially bilateral lesions) have difficulty with the task, even at very short delay intervals. The deficit may be alleviated by reducing the amount of interfering stimuli (e.g., by turning out the lights) the subject is exposed to during the delay. The WCST requires the subject to sort response cards according to one of four stimulus conditions, as defined by the designs on the cards. After ten consecutive correct sorts, the examiner shifts the required sorting principle to one of the other stimulus dimensions. The subject has to discover, through trial and error and examiner feedback, what the new sorting principle is. Patients with frontal lesions make more perseverative errors (i.e., take more trials to learn the new sorting principle) than do healthy controls or patients with similar lesions in other parts of the cortex. The frontal patients seem unable to inhibit previously rewarded responses and so they perseverate. The Stroop procedure requires the subject to name the color of the ink in which a color-word is printed. Sometimes the word will name the color of the ink (the word blue in blue ink) and sometimes the word will be the name of a different color than the ink (e.g., the word blue printed in green ink). The subject must ignore the word and name the color. Patients with frontal lesions are more vulnerable to Stroop interference than matched controls. They are unable to ignore the word on incongruent trials so they can name the ink. Taken together, the performance of patients with frontal lesions on these neuropsychological tests has revealed a consistent cognitive deficit, which is best described as an inability to resist effects of cognitive interference. The available neuropsychological evidence, then, confirms the claim that the frontal lobes are crucial for cognitive and behavioral inhibition.

FRONTAL LOBES, INHIBITION, AND EPISODIC MEMORY

PET scanning studies of memory retrieval have demonstrated significant decreases in brain activity in certain regions along with increased activity in other complementary brain sites. The decreased brain activity most likely represents neurophysiologic inhibition of those regions. Apparently, memory recall requires or involves some form of neurophysiologic inhibition. Interestingly, the brain regions most consistently implicated by PET in memory retrieval turn out to be the frontal lobes. This finding was surprising because most clinical studies of brain-damaged patients implicated medial temporal and diencephalic structures in classical forms of anteretrograde amnesia, when the patient could not learn or recall new experiences or information. Although lesion and clinical studies had also uncovered an association between frontal dysfunction and various types of memory disorder, the association was always believed to be indirect. The frontal lobes were only supposed to assist other neuroanatomical structures (e.g., the hippocampus) in doing the real work of memory. The role of the frontal lobes was restricted to helping develop strategies to facilitate encoding, storage, or retrieval. Patients with frontal lobe dysfunction, for example, do not spontaneously use categorization or chunking strategies to memorize large amounts of material. They do not efficiently encode contextual information, such as time and place, surrounding an event (or an experimental trial). They do not easily recall where or how they acquired or learned a new piece of information (source amnesia), nor can they say which piece of information, among an array of similar items, they learned most recently. Finally, patients with frontal lesions are not as proficient as controls at making judgments and predictions about their own memory abilities. All of these deficits seem to involve memory operations, rather than memory content per se, so it is not surprising that before the advent of PET scanning, the role of the frontal lobes in memory was assumed to be important but indirect and peripheral to the main action.

There were, nevertheless, some clues that the frontal lobes might be more directly involved in fundamental memory processes such as retrieval. This clue involved the following reasoning: Since conscious recollection is always associated with a sense of self who experiences the recollection, the sense of self was probably important for recall. Yet the sense of self was often impaired after frontal lesions. Personality alterations are frequently associated with orbitofrontal lesions and more subtle personality impairment is often associated with dorsolateral dysfunction. Patients with frontal lobotomies are the clearest example of impairment in the sense of self after frontal damage. These frontally impaired individuals could learn new material but they were not aware that they had learned that new material! The new material had not contacted the self structure in working memory, so the individual was unaware that he had learned anything new. If you asked him to recall a list of words you had just presented to him, he would not be able to do so. But if you asked him to just say the first words that popped into his mind, he would produce many of the words on the list. He did, in fact, retain those words, but he was unaware of that fact. Although part of the impairment of self in these disorders is undoubtedly due to disinhibition (or disconnection) of subcortical limbic (emotion) and basal ganglia (motor) sites, the role of memory impairment should not be overlooked.

Memory is clearly crucial to any enduring sense of identity and self. As Tulving and his colleagues have pointed out (Tulving, 1983; Wheeler, Stuss, & Tulving, 1997), the act of recall or remembering really involves an experience of the self. Episodic memory or conscious recollection always involves a personal consciousness and to the extent that recall leaves out this sense of self, conscious recollection will not emerge. The information will be "there," but not available on demand for the subject. From the point of view of personal well-being, personal autonomy, and self-regulation, the absence of a strong subjective experience of remembering is disastrous because memory can no longer be used to oppose excessive reliance on external stimuli to guide behavior. The internal compass is compromised and the person becomes subject to the exigencies of the moment. Personal freedom is lost. This excessive reliance on external salient stimuli is often seen in patients with frontal lobe lesions. The extreme case is called "utilization behavior," where patients slavishly pick up and utilize whatever object may capture their attention. We see similar phenomena in children with attention deficit disorder. What is missing in these and similar disorders is not just the ability to develop cognitive strategies or an "abstract attitude." Rather, what makes these disorders so malignant and harmful to the individual is loss of the subjective experience of memory—the ability to use internal, mnemonic information to avoid excessive reliance on external stimuli to guide behavior. Goldman-Rakic (1994) and others (e.g., Jacoby et al., 1989) have pointed out that one function of the frontal lobes is to call up internal mnemonic information to help guide behavior. The PET scanning studies help us to see this process as it unfolds in the brain.

PET scanning involves injection of a radio-labeled isotope into the bloodstream of a volunteer subject. The isotope attaches itself to molecules in the blood and can therefore be detected by scanners that surround the volunteer's head. Since neuronal activity requires increased blood flow, changes in regional cerebral blood flow, as detected by PET, are assumed to involve changes in neuronal activity patterns. While these blood-flow patterns are being monitored by PET, the subject engages in an experimental task devised by the experimenters to tap only a single cognitive process or function. In a review of PET studies on episodic encoding and retrieval processes, Wheeler, Stuss, and Tulving (1997; see also Nyberg, Cabeza, & Tulving, 1996) conclude that episodic retrieval is associated with an increased blood flow in the right-frontal cortex with no increased blood flow in the left-frontal cortex; episodic encoding, by contrast, is associated with the opposite pattern—increased flow in the left-frontal cortex and no increased flow in the right-frontal cortex. They call this set of findings HERA for hemispheric encoding/retrieval asymmetry. Apparently, the right-frontal involvement represents only retrieval set or retrieval mode, not actual retrieval of information itself. The right-frontal activation can be obtained even when subjects attempt but fail to retrieve items from memory. Actual episodic retrieval, or at least retrieval involving visual images, is associated with activation of posterior (parietal and occipital) cortical sites. For Wheeler and colleagues, right-frontal activation signifies establishment of a retrieval set. This idea is consistent with Bergson's selectionist model of remembering. If we assume that right frontal sites are under tonic inhibition of left-frontal and posterior sites, then relaxation of this inhibition (as evidenced by decreased blood flow to these areas) would lead to activation of right-frontal networks. According to Bergson, proliferation of an array of mental associates of the target image or cue occurs after relaxation of inhibition. This proliferation of images constitutes establishment of the retrieval set. Then the matching process proceeds via inhibition of irrelevant images. The left-frontal activation/encoding processes and its associated right-frontal inhibition would be the HERA equivalents of this matching process. Finally, selection occurs when posterior sites are activated. The role of the self in all this is, of course, obscure, but it may emerge in interaction between frontal and posterior sites. It is out of the interaction of the approach and avoidance tendencies of these two brain regions that a sense of an autonomous self emerges. It is time, in any case, to examine more closely the interactions of self and memory from the selectionist perspective. Since William James was the first to systematically explore this territory of the psyche, we will take him to be our guide.

CHAPTER 6

William James on Memory, Variability, and Consciousness

VARIABILITY IN MENTAL LIFE

Charles Darwin once asserted that without variation, selection could do nothing. Selection in the realm of the Mind implies an abundance of variable mental elements from which to select. Far from being a rather colorless blank slate upon which an environment writes, the Mind, from the selectionist point of view, is teeming with a superfluity of images, ideas, plans, and the rest. The Mind is essentially a sea of variable mental objects. The stuff of the Mind is variable, changing, heterogeneous, multiple, and diverse. We experience a welter of beliefs, attitudes, emotions, images, dispositions, anticipations, dreams, fantasies, thoughts, musical phrases, aches, pains, and the like, constantly and without let-up except, perhaps, in deep, dreamless sleep. Variability is Mind's central characteristic. This is as it should be, since variability is a fundamental property of all biological systems. It is the essential precondition for the occurrence of evolution of any kind.

William James's idea of the "stream of thought" refers to this inherent variability associated with the contents of the Mind. Like Henri Bergson, William James produced a relatively complete account of a selectionist approach to cognition. He made variability, or the "stream of thought," the centerpiece of his theoretical system and conceived virtually all other mental functions as reactions to the plenitude of mental objects in the stream. Variability forces the Mind to structure itself in ways that will reduce the overall variability. Attention, for example, operates by selection of one strand of the stream over others. Memory selectively retains a past portion of the stream. Thought involves traversing a selective portion of the stream, with the terminus or endpoint representing the conclusion or content of the thought. Personhood involves one portion of the stream appropriating other portions of the stream as vital or "self," and so on. Selection is central to all of these functions.

JAMES AND SELECTION

William James (1842–1910) is still considered by most to be America's greatest philosopher, and he was surely psychology's greatest writer. His classic text, *The Principles of Psychology* (1890), is still read for its literary beauty as well as for its scientific content. James and Bergson knew of one another's work and seemed to like one another's work. James devoted an entire chapter of *A Pluralistic Universe* (1909) to Bergson's ideas, and the two corresponded on a fairly large range of issues—though the bulk of this correspondence has never been published. In his preface to the French translation of James's *Pragmatism*, Bergson succinctly summarized James's pragmatic stance and endorsed his general approach to matters philosophical. James argued that the pursuit of ends as well as choice are the marks of human intentionality and he made choice central. "Selection is the very keel on which our mental ship is built" (1890/1981: 640). Consciousness itself was selection and in service to a personal agency: "[C]onsciousness is at all times a selecting agency. Whether we take it in the lowest sphere as sense, or in the highest sphere of intellection, we find it always doing one thing, choosing one out of several of the materials so presented to its notice, emphasizing and accentuating that and suppressing as far as possible all the rest" (1890/1981: 142).

As with Bergson's theory of remembering, selection acts to accentuate a target while suppressing its competitors. It increases the information content of the Mind. Selection, furthermore, is central to consciousness. From the simplest perception of objects in the world to the highest ethical dilemmas, James made selection and choice the laws governing mental life:

> *If now leaving the empirical combination of objects we ask how the mind proceeds rationally to connect them we find selection again to be omnipotent. . . . all reasoning depends on the ability of the mind to break up the*

totality of the phenomenon reasoned about into parts and to pick out from among these the particular one which in our given emergency may lead to the proper conclusion. . . . Reasoning is but another form of the selective activity of the Mind.

If now we pass to its aesthetic department our law is still more obvious. The artist notoriously selects his items, rejecting all tones, colors, shapes, which do not harmonize with each other and with the main purpose of his work. That unity, harmony. . . . which gives to works of art their superiority over works of nature, is wholly due to elimination. . . .

Ascending still higher, we reach the plane of ethics, where choice reigns notoriously supreme. An act has no ethical quality unless it be chosen out of several all equally possible. . . . The ethical energy par excellence has to go farther and choose which interest out of several, equally coercive, shall become supreme. . . . what he shall become is fixed by the conduct of this moment. (1890/1981: 276)

MEMORY

James's analysis of memory was succinct and linked to his conception of selection as well as the famous "stream of consciousness."

"An object which is recollected in the proper sense of that term is one which has been absent from consciousness altogether and now revives anew. It is brought back recalled, fished up, so to speak, from a reservoir in which with countless other objects it lay buries and lost from view" (1890/1981: 608). This is what may be called (though James does not use the term) the unconscious portion of the stream of memory images. Memory for James was knowledge of an event with the additional consciousness that we have thought or experienced it before. "A general feeling of the past direction in time, then, a particular date conceived as lying along that direction, and defined by its name or phenomenal contents, an event imagined as located therein and owned as part of my experience—such are the elements of every act of memory" (1890/1981: 612).

Thus, for James, remembering and personal identity are intimately linked and that link requires that memory be something more than a mere copy of an original event. Remembering is a feeling of belief in a peculiar complex mental object that includes the object, its connection or reference to pastness, and finally its connection to an experiencing self. Mental life, in fact, evolves. The class of mental objects that are selectively retained over time are called memories. This cumulative evolution of mental content constitutes a lineage of related mental images, and this lineage of images supports a bootstrapping process in mental life. Calvin (1996) and others (e.g., Campbell, 1960) have pointed out that this bootstrapping process could account for progressive increases in intelligence, but I wish to focus here on its consequences for memory and identity.

Surely, a cumulative fund of personal memories must contribute in some way to development of a personal identity. On the other hand, identity seems to be more than a mere epiphenomenon of memory. When self acts consciously it is focused, goal directed, intentional, or "a fighter for ends"—as James once said. Yet, few would disagree with the assertion that memory provides consciousness with its framework of meanings, ends, goals, and desires. Consciousness without memory appears to be an impossibility.

Yet, what precisely constitutes the nature of the relationship between memory and self is unclear. James tried to conceive the relationship as a relationship between a kind of thinking self and the "stream of thought." Though the self constitutes part of the stream of thought, it nevertheless, confronts the rest of the "stream of thought," as best it can. It strives to achieve some kind of equilibrium with the stream. It seeks to avoid being overwhelmed by it. The self experiences the stream as a vast array of variable images, as an excess, a plenitude, even, at times, a threat. It adopts a suite of strategies with which to deal with the stream. One of these strategies is memory.

PERSONAL CONSCIOUSNESS AND THE STREAM OF THOUGHT

Like Bergson, James stresses the ability of memory to confer a kind of autonomy on the rememberer from external contingencies. For James, these external contingencies (along with all other mental phenomena) are presented to us in the form of the stream of thought. James distinguished five important characteristics of the stream of thought:

1. Every thought tends to be part of a personal consciousness.

2. Within each personal consciousness, thought is always changing.
3. Within each personal consciousness, thought is sensibly continuous.
4. It always appears to deal with objects independent of itself.
5. It is interested in some parts of these objects to the exclusion of others, and welcomes or rejects—chooses from among them, in a word—all the while. (1890/1981: 220)

With regard to point number one, every thought is part of a personal consciousness, James explained. "The only states of consciousness that we naturally deal with are found in personal consciousness, minds, selves, concrete particular I's and you's" (1890/1981: 221). Each thought belongs to someone. The elementary psychic fact, therefore, is not *that* thought but *my* thought. Thoughts are always personally related, though they can be produced involuntarily. Involuntary thoughts are sometimes experienced as alien to self. They are, nevertheless, invariably imputed to some other personal consciousness. The first characteristic of the stream of conscious thought, then, is that it is invested with or possesses personhood. The question of unconscious thoughts is still debated today. James thought that these unconscious mental phenomena could themselves be referred to as subconscious selves, and so the principle still held.

This particular property of the stream seems not to be adequately appreciated by modern psychology. Instead, modern psychologists tend to treat various activities of the mind as abstract information-processing units with no particular connection to a personality. But this is a crucial oversight. The properties of the memory system, for example, change drastically if you consider memories as always connected to or in service to a particular personal consciousness. One would expect nonveridical and biased memory processing when memory processing functions to serve a personal consciousness. When personal consciousness is removed, it becomes much easier to view the memory system as a neutral, information-processing device designed to accurately record whatever happens. A selectionist approach to memory, furthermore, expects there to be, you might say, some kind of particular lineage or person with whom the cumulative memory store is associated. Functionalist approaches to memory require no personal consciousness.

With regard to point number two, that thought is constantly changing, James stressed that "[N]o state once gone can recur and be identical with what it was before" (1890/1981: 224). Change is fundamental to the variability of mental life. Each chain of thought is a succession of variable states. The stream itself is a succession of variable states. "Now we are seeing, now hearing, now reasoning, now willing, now recollecting, now expecting, now loving, now hating" (1890/1981: 224–25). Mental objects associated with these states (e.g., a particular memory), on the other hand, can seem to recur. Strictly speaking, however, nothing recurs since all of life is subject to the ravages of time. If a mental content returns a second time, it always does so with a new time tag or date and is, therefore, different than when it first appeared.

Traditional psychology should hold that this fluid property of the stream is peculiar. Functionalist memory theory, in particular, should abhor this property because it undermines the theory of the permanency of memories or of the memory trace.

James's third point, that thought is sensibly continuous, is more difficult to follow. While it seems intuitively obvious and consonant with our daily experience to note that the contents of Mind are constantly in flux, it seems less obvious that there are no gaps or divisions between mental objects. James claims that we experience the stream as "without breach, crack or division" (1890/1981: 231).

The proposition that within each personal consciousness thought feels continuous, means two things:

1. *That even where there is a time gap the consciousness after it feels as if it belonged together with the consciousness before it, as another part of the same self;*
2. *That the changes from one moment to another in the quality of the consciousness are never absolutely abrupt.* (1890/1981: 231)

When we awake, say, from general anesthesia after surgery, we may be disoriented for a brief instant. We nevertheless soon regain our bearings and the stream of thought flows just as it had before the interruption. The stream is experienced as a common whole and the natural name for it is myself, I, or me. It is characterized by feelings of warmth, intimacy, and immediacy, and these feelings contribute to the experience of the stream as both personal and continuous.

There are, nevertheless, contrasts in the quality of successive segments in the stream of thought. The rate of change among components of the stream may differ. If the rate of change is rapid, we spend less time contemplating a particular mental object and more time noticing transitions between mental objects. James called these points of transition the transitive parts of the stream. He contrasted these with the substantive parts of the stream. "It then appears that the main end of our thinking is at all times the attainment of some other substantive part than the one from which we have just been dislodged. And we may say that the main use of the transitive parts is to lead us from one substantive conclusion to another" (1890/1981: 236). James pointed out that to ignore transitive parts of the stream is to ignore the internal relations of mental objects. Internal relations of objects are, of course, critical for knowledge of these objects.

One of the most interesting aspects of the stream is what James called the halo or penumbra of images that surround a target image and confer significance on it.

> *Every definite image in the mind is steeped and dyed in the free water that flows around it. With it goes the sense of its relations, near and remote, the dying echo of whence it came to us, the dawning sense of whither it is to lead. The significance, the value, of the image is all in this halo or penumbra that surrounds and escorts it—or rather that is fused into one with it and has become bone of its bone and flesh of its flesh; leaving it, it is true, an image of the same thing it was before, but making it an image of that thing newly taken and freshly understood.* (1890/1981: 246)

The halo of images around a target image might be considered equivalent to the set of images generated during the proliferation phase of a given cycle of Bergsonian remembering. During the generative proliferation phase, the rememberer generates a huge array of associates in response to the "cue" image. These associates help to establish a cue-related retrieval context which can assist in the selection-recall process. The associates, in other words, confer significance on the cue and its target. Tulving (1983), following Semon (1909/1923), called a retrieval process like this "synergistic ecphory," as the combined cue-target information leads to conscious recollection.

Philosophers have pointed out that what makes an experience conscious is its dependence on "qualia." Qualia are experiential properties of mental states. They refer to the ways in which it feels to be in pain or to hear or to see or to feel. If we could explain how the physical world (e.g., the brain) gives rise to qualia or experiential "feels," we would be one step closer to explaining the nature of consciousness—at least the basic, phenomenal awareness every sentient creature probably experiences. What gives rise to specific qualia for a given act or experience? To a first approximation, it must be the activation of a set of associates to a target image or mental content. Qualia on this account refer to the halo or penumbra of images that confer significance on a selected or target image in the stream of consciousness. If we assume a kind of network architecture for the brain/Mind and we further assume that every image (or node in the network) maintains connections with other images in other parts of the network (representing other brain sites), activation of associates of a target image will recruit images from other brain systems such as affective and motor sites as proliferation proceeds. As the recruitment process unfolds, affective, mnemonic, and other associations will emerge, thus conferring some kind of experiential significance on the event. When the event is remembering, conscious recollection will emerge. When the event is perceiving, conscious seeing or hearing or touching will emerge. From the point of view of the stream of thought, the generation of associates (the halo or penumbra) to a target image constitutes a mental traversal of one portion of the stream to another portion of the stream.

We all have an intuition of this feeling of tendency in our thought, this feeling of where our thought is going. The feeling or intuition comes out of the halo of relations each thought is bathed in. We can almost feel ourselves traversing the transitive portions of the stream in pursuit of some substantive conclusion. Knowledge about a thing is derived from these traversals of the stream as well as this intuition concerning its relations. The meaning of a thought is its conclusion—its endpoint after a traversal.

In any case, this property of the stream, the fact that it is sensibly continuous, was, according to Bergson, a particularly important characteristic of that portion of the stream composed of memory images. For Bergson, duration was devoid of spatial properties. Memory images were not separated by spatial distances but by qualitative differences. If Bergson is correct about this claim, the consequences for memory theory could be immense. If space was not essential to memories, then the storage metaphor collapses. Similarly, if memories are "contained" in a nonspatial realm, then communication of memory images between two personal consciousnesses should obey laws different from those of verbal communication. Finally, if memory is non-spatial, then the identity or identities that memory supports should not be reducible to one particular, spatially organized "container" like a brain. Unfortunately, none of these implications have received any substantial scientific investigation. One possible exception might be the case of psi abilities like telepathy, but such abilities are controversial.

Point four, thought deals with objects independent of itself, is another way of saying that thought exhibits intentionality or knowing. In terms of the stream, intentional thought leads to a conclusion at a substantive segment of the stream, and this conclusion is its object. To cogitate, to know, is to traverse a portion of the stream and reach a conclusion. But who does the traversing? Thought or self? Once again, we return to the issue of a personal consciousness.

Point five raises the issue of selective attention. Some parts of the stream are selected, chosen, emphasized or ignored over others. James paints a picture of a progressive filtering of sensations or information, which come to us from the periphery. First, we are bombarded with stimuli, we then select salient distinctions/attributes from these stimuli, then these salient aspects are subjected to further selection until a quite definite percept or concept is produced. Rationality itself depends on selection: "If now leaving the empirical combination of objects, we ask how the mind proceeds rationally to connect them, we find selection again to be omnipotent. . . . [A]ll Reasoning depends on the ability of the mind to break up the totality of the phenomenon reasoned about, into parts, and to pick out from among these the particular one which, in our given emergency, may lead to the proper conclusion" (1890/1981: 276).

James's classic treatment of the stream of thought can be summarized as follows: We encounter the stream as composed of mental objects whose origins lie in a personal consciousness. These mental objects include transitive and substantive portions of the stream, as well as feelings of tendency and a variety of mental states and their products. Thinking involves a traversal from one substantive portion of the stream via the transitive portions (made of internal relations of mental objects) to arrive at a new substantive portion of the stream. The stream is experienced as a smooth continuous whole that flows. The stream's most salient property is incessant change and variety associated with a personal consciousness.

Bergson's notion of duration seems to be related to James's ideas on the stream of thought. Duration is composed of a multiplicity of nonspatial images or movements. One sees, for example, a multiplicity of vibrations or wavelengths as red. Pure duration is the lived experience of time.

Bergson compares duration to a melody. A melody can be appreciated as a melody or as a set of discrete notes. Time and duration can also be appreciated holistically or as a discontinuous, spatially conceived line, which projects backwards and forwards from a present point. In duration, however, memory images interconnect and interpenetrate; that is, they flow. Duration is characterized by heterogeneity, multiplicity, plurality, qualitative differences, and so on. James and Bergson thus apparently agreed on the fundamental qualities of the stream of thought—though Bergson emphasized its memory-related aspects.

The stream evidences other properties as well, not discussed by James. These are its limitlessness, heterogeneity, and limited reversibility. The stream is, as James noted, in constant flux. We notice that our experiences, perceptions, and thoughts are always in transition, in flux. The flow is involuntary and independent of any internal circumstances, save perhaps deep sleep or deep anesthesia. The flow of thought, furthermore, seems unconstrained by informational types. We can spontaneously imagine many types of information at once. All of these factors lend to the stream an impression of indestructibility or limitlessness. There is no end to the outpouring of mental images until death. A wealth of mental images flow continuously and constantly through our minds whether we like it or not until the day we die.

The flow of thought, therefore, seems limitless. When we add to this the fact that the constant flow is made up of heterogeneous images, we are humbled. We envision an inexhaustible flow of an infinite variety of images—from goal-directed thought to involuntary fantasy and much else besides. Even if we take only one subdivision of the stream, say, memory images, the wealth and variety of types is staggering.

Finally, the stream exhibits limited reversibility—that is, certain portions of the stream can be re-cognized via memory but this represents only a very limited portion of the stream. Nevertheless, it is this portion of the stream that makes all the difference for us humans, since it gives rise to personal knowledge and identity.

Despite the bewildering variety of mental phenomena presented to us in the stream of thought, we can begin to classify it in terms of its sources. The stream of mental occurrences is composed of two major subdivisions: the internal and the external subdivisions. The internal stream comprises images drawn from the inner life of the individual. These images include visceral, proprioceptive sensations, memories, beliefs, ideas, wishes, dreams, stories, schemas, models, thoughts, plans, strategies, and the like. This inner stream of images may be further subdivided into voluntary or self-generated images and involuntary or automatic images. The external stream is composed of perceptions, affordances, group phenomena, cultural phenomena, and other complex perceptual imagery. While these broad subdivisions are useful starting points for a study of the properties of the stream, they do not account for crossover images like perception of sensory pain and the like. Nevertheless, the common-sense division of the stream into inner verse outer realms will do for now.

Pope and Singer (1977) suggested that individuals could develop a cognitive or perceptual set or bias toward either the internal or external components of the stream. The mental set would be facilitated or implemented by attentional processes, current concerns, current affect, and whether or not the current environment was predictable or surprising. Singer and colleagues (Singer, 1993) developed a signal-detection paradigm to explore the properties of the stream. Subjects sit in a soundproof room and are required to attend to a signal-detection task. They may, for example, be required to press a button every time they think they detect a faint red light. Periodically, subjects are queried (via intercom) about whether any task-unrelated images and thoughts (TUITs) occurred during the ongoing signal-detection task.

Singer and colleagues find that TUITs occur at least half the time when subjects are achieving very high accuracy rates on the detection task. Most TUITs turn out to be memories, wishes, fantasies, and other thoughts far removed from the immediate setting. People in Singer's experiments are surprised at how much time they spend daydreaming. Singer found some interesting modality-specific interference effects. When he used auditory signals for the detection task, subjects were more likely to report visual images and when he used a visual signal, subjects reported a greater number of auditory images. Individuals who are known by self-report to be frequent daydreamers are much more likely to report TUITs than individuals who report that they are not frequent daydreamers. High levels of emotion may increase the number of TUITs reported by subjects.

Klinger (1978) has distinguished two types of thoughts in the ongoing stream: operant and respondent. Operant thought, like Freud's secondary-process thinking, is effortful, conscious, volitional, and instrumental. It is directed toward a specific end, like the solution of a specific problem.

Respondent thought, on the other hand, is like Freud's primary-process thought. It is nonvolitional, effortless, and includes all other types of thought. Klinger finds that the largest share of daydreams focus on current concerns—though not necessarily on operant thought. Operant images must surely reflect our propensity to act or operate upon the world. This brings up the issue of technical imagery—another very important part of the stream.

The philosopher Dessauer (1983) referred to this realm of technical imagery as the fourth realm. The other three realms were described by Kant and might be construed as other parts of the stream. The first realm refers to sensory experience. The a priori categories of intuition, time, and space give rise to this experience and the law of causality, and makes science possible. Kant asserted that the "thing-in-itself," outside the categories of space and time is unknowable by pure reason. Kant's first critique (of pure reason—see translation by N. Kemp Smith, 1929) was devoted to these issues. His second critique (of practical Reason—see translation by L.W. Beck, 1956) was devoted to the second realm of moral law. Moral law is based on an unconditioned ought and ideas of religion, God, and the absolute. Moral intuition allows us to move beyond mere appearance and phenomena and encounter the thing in itself. The third realm, according to Dessauer, is the realm of feeling or aesthetic. The beautiful is freedom (second realm) in the world of appearance and causality (the first realm). The fourth realm is the realm of technical imagery and action—a subclass, we might conclude, of Klinger's operant thoughts. Dessauer asserted that the fourth realm is to be understood as the struggle to make the categories (of the first realm) conform to the thing-in-itself—to realize (through human work) in the world of appearance the thing in itself. When forms from the fourth realm are realized in the world (as mechanical devices), a great deal of power is unleashed. Thus, the effectiveness of technology.

One final property of the stream deserves consideration—individual contributions to it and its relation to a suprapersonal common or collective stream. Here we are in complete darkness. If we take the model of individual relations to a common culture as appropriate for the question of individual relations to a common stream, then some progress is possible. I will return to the question of culture in a later chapter. What about individual relations to a local, ongoing group—stream of thought? Here I suspect affect reigns supreme. Groups are moved mostly by emotions. Emotions are communicated nonconsciously and often irrationally. Take the simplest case of a dyad. Although verbal communication is the norm, unspoken affective communication is probably more important for determining real exchange of information between two persons. The thousand cues associated with emotion are fed into the stream and transmitted from one person to the other in a constant, ongoing interchange of unspoken thought. Here I only wish to register this fact, not to analyze it. Affect is an important part of the stream and contributes mightily to group communications.

As James argued (and as selectionist theory predicts), variability of the stream of thought is the central fact of the mental life of human beings. All individuals have to come to terms with it if they are to be enriched by its contents. Interaction of the self with any portion of the stream occurs along two broad poles: the external or public realm and the internal or private realm. Personal autonomy is the result of achieving a balance between these two poles, but this is the topic for the next chapter.

CHAPTER 7

The Theater of Simultaneous Possibilities

SELF AND MEMORY

We now enter into exceptionally difficult territory—the conceptual territory surrounding theories of the self, identity, ego, consciousness and personality, in relation to memory. James introduced his treatment of the self in *The Principles of Psychology* (1890/1981) with the concluding paragraph of the chapter on the stream of thought:

> *[T]he mind is at every stage a theater of simultaneous possibilities. Consciousness consists in the comparison of these with each other, the selection of some, and the suppression of the rest. . . . But in my mind and your mind the rejected portions and the selected portions of the original world-stuff are to a great extent the same. The human race as a whole largely agrees as to what it shall notice and name, and what not. . . . There is however one entirely extraordinary case in which no two men are ever known to choose alike. One great splitting of the whole universe into two halves is made by each of us; and for each of us almost all of the interest attaches to one of the halves. . . . When I say that we all call the two halves by the same names, and that those names are "me" and "not me" respectively, it will at once be seen what I mean.* (1890/1981: 277–278)

He summed up his chapter on the consciousness of self this way:

The consciousness of self involves a stream of thought, each part of which as "I" can 1) remember those which went before, and know the things

> *they knew; and 2) emphasize and care paramountly for certain ones among them as 'me' and appropriate to these the rest. . . . Whatever remembered past feelings resemble this present feeling are deemed to belong to the same me with I . . . . This me is an empirical aggregate of things objectively known. The I which knows them. . . . is a Thought, at each moment different from that of the last moment, but appropriative of the latter, together with all that the latter called its own. . . .* (1890/1981: 378–379)

James, after reviewing such clinical phenomena as multiple personality, hypnosis, and trance states, went on to assert that the same brain may subserve many conscious selves. These coexisting selves later became important in twentieth-century psychology when attempts were made to explain akrasia and self-deception.

James ascribed the sense of personal identity or sameness over time to the action of one bit of thought from the stream judging other bits of thought, perhaps in adjacent portions of the stream, as belonging together. "This sort of bringing of things together into the object of a single judgment is of course essential to all thinking" (1890/1981: 315). It is otherwise known as recollection, from the Latin *colligere*—to bring or gather together. If, by judging bits of the stream as belonging together, James was referring to a kind of matching process that constitutes selection in the realm of memory, then we can say that continuity of identity over time emerges naturally from selectionist mechanisms operative in the realm of memory. Growth in memory is driven by repeated selection cycles and cumulative retention of selected material. Growth, furthermore, is internally driven, since selection is made from an array of internally generated images (with interaction from external cues). Past images/memories are used in arriving at current memories; thus continuity with the past is ensured. Selectionist evolution is therefore dialectical in the classical sense of growth that emerges out of a synthesis of contrary elements, such as past and present elements. But what about the content or character of identity? Identity is more than just "continuity over time." There is a character who remains the same and yet changes over time. There is a personality that undergoes change and grows.

SELECTION AND SELF

For James, content or character was "acquired" via choice or selection from an array of possible characters or selves:

> *[Many] . . . different characters may conceivably at the outset of life be possible to a man. But to make any one of them actual, the rest must more or less be suppressed. So the seeker of the truest, strongest, deepest self must review the list carefully, and pick out the one on which to stake his salvation. All other selves thereupon become unreal, but the fortunes of this self are real. Its failures are real failures, its triumphs real triumphs, carrying shame and gladness with them. This is as strong an example as there is of that selective industry of mind on which I insisted some pages back. . . . Our thought, incessantly deciding which ones for it shall be realities, here chooses one of many possible selves or characters, and forthwith reckons it no shame to fail in any of those not adopted expressly as its own.* (1890/1981: 295–296)

James seems to be suggesting that something like ready-made characters or self-concepts are among the objects which can be selected from the stream. These self-concepts are mental entities that support development of character. What are the properties of this complex mental representation we call the self-concept? It must include both abstract information about a person's characteristics and concrete information about a person's experiences, thoughts, and actions. Kihlstrom and colleagues (1988) have investigated the formal properties of the self-concept as represented in working memory. They argue that this representation is a declarative knowledge structure which can be represented as a graph structure, with nodes representing concepts and associative links representing the relations between them. The nodes can be thought of as propositions or predicate-argument structures, which encode both semantic and autobiographical memory content. A node representing the self, for example, would be linked, in such a graph structure, to propositions like the following: likes Stravinsky, lives in Boston, loves Mary, and so on. It would also be linked to episodic information like this: helped an old man across the street on Monday, mentioned new book to Mary yesterday, and so on. While some memories would support an overall self-assessment, such as "I am a kind person," others would be inconsistent with such a conclusion, generating conflict. Undesirable memories may not be directly linked to the self node at all, and thus would be less accessible to the conscious self. In addition to these knowledge structures that are linked to the self node, we find that other parts of the mental system, such as those regulating drives, affects, goals, and preferences, are also fed into the self network.

Unlike other mental representations, the self has privileged access to immense energetic sources in the organism. When it is recalled that the self also retains privileged access to perceptual circuits (i.e., external information), then it becomes clear that the self is the organism's preferred way of dealing with the vast imagery of the stream. Thus, the self might be considered the most important system in the domain of mental life. All of the person's mental and emotional energies are tapped and perhaps regulated by the self structure. The person's normal functioning, therefore, depends on the integrity of the self structure. No wonder it is exceedingly complex.

HIERARCHY OF SELVES

James believed that we are presented (in the stream of thought) with a vast number of these ready-made self-concepts and that, over time, we select out and cumulatively store a range of different selves. These selected selves presumably are then developed to a greater or lesser degree in interaction with the environment. They are, in any case, ordered into a hierarchical scale "with the bodily self at the bottom, the spiritual self at top, and the extra corporeal material selves and the various social selves between. . . ." (1890/1981: 299). If selectionist mechanisms are operating at the level of self or self-concepts, then such a hierarchical ordering of selves is inevitable. If self-concepts compete against one another for selection and if selective retention is cumulative, then old selves will not be discarded. Rather, they will contribute to development of new or current selves. Given the cumulative nature of selectionist systems in general, selves higher in the hierarchy should be partly synthesized out of selves lower in the hierarchy and thus be able to demonstrate some level of access to the information all other selves possess. Selves lower in the hierarchy, on the other hand, should not be able to demonstrate knowledge of all of the information the dominant self possesses.

SELF-AWARENESS

James's selectionist account of the self and personality development carries with it implications for a theory of self-consciousness. Like any other selectionist process (e.g., memory), the selves that win out in the competition will constitute a cumulative store of selves from which new possibilities will be generated for each new selection cycle. The added complication here is that each of the selves has the capacity for self-awareness, depending on where in the hierarchy of selves a target self occurs. Consciousness, as James says, consists in the comparison of these possibilities against one another, the selection of some, and the suppression of the rest. Just as phenomenal awareness or the "feel," (i.e., qualia) of things arises out of the generation of memory associates to a target sensation or memory—and the ways in which these associates (a) recruit affective and sensory systems, thereby (b) contextualizing the target image—so, too, does self-consciousness arise out of the generation and evaluation of an array of possible selves. This array of possible selves will help to contextualize the current target self so that its many-layered significances for the person's possibilities will emerge. The contextualization process puts the spotlight on the target self, and the target self undergoes a kind of evaluation before it is selected as the current or topmost self. When a given self is selected, it moves to the top of the hierarchy and becomes the self we identify with. Selves lower in the hierarchy, however, if not continually inhibited, continue to compete for mastery with the currently dominant self, and this dynamic creates conditions for dissociations in self-awareness. The entire Jamesian account of self and consciousness seems to imply that consciousness could easily be fractionated across several centers in the hierarchy. These centers would probably involve subordinate selves or current selves that are competing for dominance in the hierarchy.

FRACTIONATION OF SELF-AWARENESS

There are many examples of the fractionation of self-awareness in both everyday life and after injury or trauma. Some examples from the neurology clinic appear to involve disinhibition of "lower" selves in the hierarchy, while others appear to involve dissociations at a higher level of self-awareness. These latter cases usually involve "not knowing that you know" and sometimes outright denial of otherwise patently true facts. According to McGlynn and Schacter (1989), outright denial of illness occurs most frequently in the following syndromes: hemiplegia, Anton's syndrome, hemianopia, amnesia, head injury, dementia, and aphasia. The general term for unawareness or denial of illness syndromes is anosognosia. In this family of disorders, the patient will often deny any illness despite overwhelming evidence to the contrary.

DISINHIBITION OF SUBORDINATE SELVES

Early evidence for disinhibition of lower-level selves came from a 1868 report by the physician Harlow on his patient Phineas Gage. Gage, a railway worker, survived an explosion that blasted an iron bar (about four feet long and one inch wide) through his frontal lobes. After recovering from the accident, Gage's personality changed. He became irascible, unruly, and his behavior was inappropriate (Blumer & Benson, 1975). The damage had mostly been in the orbital frontal region of Gage's frontal lobes. It has become clear since Gage, and from the study of other patients with orbitofrontal lesions, that this area of the frontal lobes is crucial for regulation of drives, emotions, personal memories, and other factors related to personality. Some of the most common symptoms associated with orbitofrontal lesions include irritability, inappropriate sexual behaviors, emotional lability, occasional outbursts of rage, and inability to maintain focus and attention on the task at hand. Loss of essential qualities of personality and symptoms associated with orbital lesions also occur in the late stages of Huntington's disease and in Korsakoff syndrome. Huntington's disease is a motor disorder associated with atrophy of the caudate nuclei and the putamen in the basal ganglia. The basal ganglia are a set of cell groups that are densely interconnected with the frontal lobes. In the late stages of the disease, atrophy spreads from the caudate nucleus up into the frontal and temporal lobes; patients sometimes begin to demonstrate many of the disinhibited personality changes characteristic of orbitalfrontal lesions. Evidence of disinhibition of subordinate selves might be found in cases where a patient reports the experience of an unseen person or presence.

Critchley has discussed the neurology of the experience of a "presence" (Critchley, 1979). In some cases the presence is not yet localized to a place within the remembering subject. Rather, it is displaced to a point outside the subject. Occasionally, localization of the presence fluctuates (within-without) or is in the process of being localized within the subject. For example, one patient declared that the left half of his body did not belong to him. Sometimes, while walking, he was seized with the conviction that he was being followed by someone, namely his double. Whenever this idea developed, he would lose the feeling of strangeness that affected the left half of his body.

Then there are cases where the presence is well constructed enough to carry an identity and to be localized within the subject and yet still not be identified as the dominant self. A well-known poet told Critchley that during a childhood illness he developed a belief that someone had entered his body and begun to share his pains and struggles. Sometimes in cases like this, the subject is just on the verge of realizing that the visitant is self. One patient, for example, would often wake up in the middle of the night with the strong feeling that someone whom she knew quite well was in the room with her. Sometimes she would realize that this person was none other than herself or her double. Sleep disturbance seems to be a common theme in many of these personality alterations. Many patients with multiple personality disorder first experience the alternate self upon awakening. They go to sleep their usual self but wake up in a different personality.

Alien-hand syndrome is a particularly dramatic form of disinhibition of a subordinate self (Goldberg, 1987). Here the patient has sustained a lesion that functionally disconnects the right and left hemispheres. The patient experiences one hand as alien—as belonging to another self. The alien hand appears to have its own purposes and intentions. Sometimes, one hand will be buttoning up a shirt while the other hand will be unbuttoning the shirt, or one hand will pick up the phone to dial while the other hand will attempt to hang up the phone. In rare (but documented) cases, the patient will awake with the alien hand at his throat attempting to strangle him!

There are other accounts of disinhibition of subordinate selves, such as in multiple personality disorder, hypnosis, and doppelganger effects associated with temporal lobe epilepsy. But it is not necessary to detail these rather exotic instances of disinhibition when the phenomenon is known to occur in more familiar circumstances. Take the example of putting an otherwise normal eighteen-year-old boy into the middle of a war zone where he is being shot at, bombed, and so on. If you hand him a machine gun and then tell him that enemy villages need to be destroyed in order for them to be saved, that boy may participate in the destruction of the village—along with all of its inhabitants including noncombatant men, women, and children. Subordinate selves are a reality, and usually are regulated by selves higher in the hierarchy.

NOT KNOWING THAT YOU KNOW

Dissociations in awareness are hard to explain if you assume a single unified consciousness. Yet dissociations in awareness are ubiquitous in neuropsychology. In most of the well-studied amnesic syndromes, for example, patients typically evidence normal intelligence levels, normal language and perceptual functioning, and relatively normal psychiatric functioning. Yet they are unable to normally acquire many types of new information. In Korsakoff syndrome, insight into the memory problem is often minimal or lacking entirely. When patients exhibit some awareness of their memory problems, they often minimize the severity of the deficit or they confabulate an excuse as to why their memory is poor. Sometimes the confabulations are so outlandish and unlikely as to be bizarre. Patients are often unaware of the inappropriateness or bizarreness of some of their confabulated stories. Confabulation tends to decrease as awareness of deficit increases. Severe confabulators tend to respond quickly and unthinkingly to questions that require memory, while mild confabulators show a greater tendency to self-correct their verbal responses, indicating a greater degree of awareness. When patients are aware of their memory disorder, they are more likely to decline to answer a question or admit that they do not know the answer, as compared to patients who lack awareness of their memory disorder.

Even when patients exhibit some awareness of their memory problems, they are often still unaware of various aspects of their memory disabilities. If, for example, a list of words is repeatedly presented to a group of amnesic patients with instructions to remember the list and to learn the words, the patients will eventually learn some portion of the list. However, they will not realize that they have, in fact, learned them. When asked to produce the words on the list they will answer; "What list?" We can demonstrate that the learning occurred by providing the patients with cues to help them retrieve the words or by simply asking them to write down the first words that come to mind. Patients will usually write down words from the list, thus demonstrating that the words were learned. They had learned some new words, but were not aware that the learning occurred. Similarly, you can teach amnesic patients a few simple melodies on a piano. The teaching is laborious and takes a very long time, but eventually some learning occurs. Patients can tap out the melody they studied. If you ask them to play a piece—any piece on the piano—they will reply that they know none. If you sit them down and place their hands on the keyboard, they will tap out the melody they learned. If you ask them why they played that melody, they will claim that they just improvised it on the spot. They have no awareness that they know the melody—no memory of acquiring the melody. The patients don't know that they know.

Dissociations in awareness are well-documented among comissurotomy patients. These are patients who have had the corpus callosum sectioned to prevent the spread of seizure activity from one hemisphere to another. Marks (1986) provided a critical assessment of claims suggesting that experimental findings with these patients demonstrated that the split-brain patient has two minds or two persons in one body. A key experiment, much discussed by Marks, is the following: The phrase "key ring" is flashed on a screen for a tenth of a second, and patients are required to report what they saw. They respond "ring." If on the other hand, patients are asked to retrieve with their left hand what they saw from an array of items, they will retrieve a key and not retrieve a ring. The patients' verbal response (controlled by the left hemisphere), therefore, indicates an awarenes of "ring" but not "key," while the patients' manual, left-handed response (controlled by the right hemisphere) indicates awareness of "key" but not "ring." As Marks points out, someone saw "key" and someone different saw "ring" but no one saw "key ring."

MANY SELVES AND SELF-DECEPTION

Fractionation of awareness also occurs in everyday life in the form of self-deception. In this case, dissociations in awareness may more clearly reveal the work of subordinate personalities. Although we usually experience ourselves as a single unified identity, there are times when we do not—as when we are self-deceived over a serious personal issue or when lapses in brain or mental functioning occur. There is, in fact, a substantial body of evidence suggesting that each of us contains several selves and that the simple-minded formulation—one brain–one person—is misleading. As the Jamesian hierarchy of selves would predict, we appear to possess subpersonalities, or at least subsystems of some kind, which exhibit intentional states that sometimes conflict with "our" current mental state. One of the clearest examples of this kind of dissociation in self-awareness is self-deception.

The Renaissance essayist Montaigne observed that lying and self-deception were intimately interdependent, but that self-deception ultimately depended on memory. One could not self-deceive unless one edited out or sequestered those memories that carried the truth of the matter. Philosophers have pointed out that for subjects to be self-deceived, they must simultaneously hold two contradictory beliefs (say: *p* and not *p*). Since this is logically impossible, some philosophers have claimed that self-deception is impossible! In the strict sense, they may be right: One person cannot hold to two beliefs that are in contradiction. But one brain can. Two persons, in one brain, can hold mutually exclusive beliefs, and since the Jamesian hierarchy of selves claims that we are composed of more than one person, self-deception is not entirely problematic from James's selectionist perspective.

Nevertheless, the problem of self-deception is not entirely satisfactorily explained from this perspective either. If we suppose that an unwanted belief is suppressed by the topmost self, how do we know when it will be adopted by some subordinate self? Just how does a subordinate self acquire the rejected information? Conversely, how does the dominant self reject the information? Treating information in a motivationally biased way may involve anything from mild denial of the facts to outright suppression or sequestration of these facts away from consciousness. This strong form of self-deception appears to be ubiquitous in everyday life and is amazingly easy to demonstrate in the lab.

Quattrone and Tversky (1984) required subjects to immerse their arm into a tub of circulating cold water for as long as they could. After obtaining baseline measures of cold-pressor tolerance, one group of instructed subjects were told that increased tolerance of the cold-water pain indicated a healthy heart. On the second cold-pressor trial, these subjects increased their tolerance of the cold water pain but denied trying to do so. Control uninstructed subject's tolerance levels did not change from baseline. Instructed subjects simply did not believe that they tried to increase their tolerance on the second trial despite their higher tolerance scores. All people would probably like to believe that they have a healthy heart. In order to support this belief, instructed subjects purposefully tried to shift their tolerance levels and then had to make themselves believe that they did not purposefully shift their tolerance levels. If they let themselves become aware that they did attempt to increase their tolerance then they would have to admit to themselves that they may not have a healthy heart.

The evolutionary biologist Trivers (1985) discussed five mechanisms of human self-deception (also noted by Greenwald, 1981): benefecttance, exaggeration, the illusion of consistency, the perception of relationships, and perceptual defense and vigilance. He argued that we all have a tendency to try to represent ourselves as being beneficial and effective at the same time, to exaggerate our competencies in retelling, to rewrite past experiences to make them consistent with present conditions, to keep a cumulative tally of favors and misdeeds of others with whom we are in relationship, and to see what we wish to see and in so doing to remember such perceptions in a more positive light than reality would allow. All of these strategies of self-deception may depend on extensive editing operations in autobiographical memory to support them. To see ourselves, for example, as consistently competent, beneficial, and effective, we need to selectively access memories that yield those judgments. To dredge up memories of our incompetences would hardly do.

Trivers (1985) suggested that self-deception may have arisen out of a cognitive arms race between people who try to deceive others and those others whose interests lie in detecting deception. The better humans became at deceiving, the greater the selective pressures on humans to detect deception. To consciously deceive, however, involves a certain amount of stress: increased autonomic activity, sweaty palms, nervous agitation, various types of facial cues, and so on. It is very likely impossible for one self to consciously control all of these autonomic indices of deception. If the deceiver, however, was unaware of his deceptive communications, then all of the associated autonomic indices would be dampened and thus would be more difficult to detect. The best liars are those who actually believe their own falsehoods. But to believe one's own falsehoods may require active suppression of unwanted information, and this information has to go somewhere. If James was correct in supposing a hierarchy of competing selves in the typical human psyche, then one of these competitors may appropriate that information and use it for its own (competitive) purposes.

Whatever its origins, self-deception is ubiquitous in everyday life and it reminds us that we are not alone and, indeed, are not masters in our own house. Although the available evidence for the Jamesian hierarchy of selves is not convincing, the hierarchy is consistent with a selectionist approach to identity, personhood, and self. The Mind chooses from an array of possible selves, which, in turn, are generated from a cumulative store of such "selves." James, however, neglected to provide an explanation of how culture influences this stock of selves from which we choose.

SELF AND SOCIETY

Although I cannot do justice to the ways in which selves are constructed by and in social interaction, it is clear that the two mutually construct one another: Selves give rise to social groups, and cultures give rise to selves. The division of labor in a given society imposes restrictions on the range of possible roles or identities available to a society's inhabitants. Thus, it places an initial constraint on the range of socially acceptable realities for a given culture. Within that initial range of social roles another range of possible selves or characters becomes available to the developing person. This range of character types are often depicted in the theater or on stage, in masks, in stories and folktales, in caricatures, in folkloric typologies, and in the arts.

Historical and ideological conflict often involves a struggle over the range and content of the possible selves available from which to choose. Struggle over the division of labor is inextricably bound up with the struggle over the range of possible role types that will be available. History may deepen the content of some identities and thereby enrich the choices. In preliterate societies, myth provided choices. Thus, if individuals could not accommodate themselves to the current range of choices, they could turn to oral memories of the culture for models of alternative personality types. If people chose one of those, their current personality type would be correspondingly deepened (or complexified).

There is a tension in this dialectic between memory-dependent identities, shaped by the division of labor, and the need to select from the array of possibilities. If you increase the range of possible identities (liberal solution), you get a corresponding increase in freedom for individuals, but you also get an increase in the amount of disorder associated with the proliferation of identities. If, on the other hand, you restrict the range of possible identities (conservative solution), you increase order at the expense of freedom. The modern solution of preserving a rigid set of identities for the workplace and a larger, less rigid set of identities for nonwork arenas is an unstable compromise at best, which cannot last. The split in consciousness required by such an arrangement is too taxing for most healthy adults. The best solution would be to make the range of possible selves associated with the division of labor consonant with the set of possibilities associated with non–work-related selves. But that solution requires a fundamental restructuring of the modern industrial division of labor in favor of radical democratic control over what is produced in the society.

CHAPTER 8

The Stream of Thought and Self-Regulation

SELF AND THE STREAM OF THOUGHT

From the point of view of the dominant self, the stream of thought might be experienced as an excess, a superfluity of mental content, a superabundance of imagery and stimuli, which, therefore, needs to be reduced, dissipated, or "handled" somehow. Mental economy and balance in the Jamesian hierarchy of selves is simply not possible without some form of regulation of this hierarchy vis a vis the stream.

The prerequisite for interacting profitably with the stream of thought is the development of inhibitory powers. If inhibitory powers are compromised, one becomes subject to the exigencies of the stream. We can be plagued, for example, by unwanted tunes or melodies or thoughts that we just can't seem to shake, and we sometimes find it difficult to concentrate on the task at hand due to a lot of interfering or even intrusive thoughts. In the extreme example of obsessive-compulsive disorder, patients are unable to keep unwanted and disturbing thoughts from consciousness, thereby seriously impairing their daily functioning. That most of us experience the everyday failures of cognitive inhibition as minor annoyances that are rapidly corrected suggests that we handle our interactions with the stream of thought relatively well and effortlessly. This effortless self-regulation of our mental lives speaks to the silent but beneficent effects of selection. Our primary tool for navigating and benefiting from the stream of thought is selective processing. Selective processing allows us to ignore or filter out irrelevant information. It allows us to focus on what is helpful or useful and suppress what is not useful or painful. But selective processing depends, of course, on the ability to inhibit irrelevant or interfering stimuli or portions of the stream. Thus, inhibitory powers and selective processing are critical for normal cognitive functioning and self-regulation.

TWO STANCES

Successful self-regulation involves a balance between efforts to increase the signal regardless of background noise levels, and efforts to directly reduce noise levels in the hopes that the signal will then stand out. In terms of the person's relation to the stream of thought, these strategies imply two complementary dispositions toward the stream. If a individual's preferred self-regulatory strategy is to diminish the noise, they will tend to adopt an instrumental and repressive stance toward the stream. If, on the other hand, their preferred strategy is to enhance the signal regardless of its effect on background noise, then they might adopt a more open attitude toward the components of the stream.

We might classify the first group (i.e., those who adopt an instrumental stance toward the stream) as repressive or attenuation types (rather than augmenters), with thick boundaries (Hartmann, 1991) and extroverted personalities. Those, on the other hand, who adopt an openness to the stream of mental occurrences would be augmenters, daydreamers, and introverts with thin boundaries.

These two major forms of relationship with the stream may also be understood as strategies for regulating the internal hierarchy of selves. A dominant self may be open to dialogue with sub-personalities or may be closed to any such dialogue. Closed subjects (those who adopt the instrumental attitude) are averse to the noisiness and wild variability of the stream of thought and of competing internal voices. They withdraw from the noise and avoid it. They are thus less interested in the psychic realm than are open subjects. For those subjects who are well disposed toward the content of the stream, a preoccupation with the inner world and its various characters slowly takes precedence in the person's mental economy. These subjects approach the stream of mental occurrences rather than avoiding it. They are interested in finding the signal, the "key," the riches that inhere in the inner life.

When an overemphasis is placed on the contents of the stream in an effort to locate the signal, the individual rapidly becomes inner focused and self-absorbed. External stimuli are neglected, and internal stimuli are given too much weight or significance. When, on the other hand, too much emphasis is placed on reducing inner stream–related occurrences (or noise), external stimuli are given too much weight. The person's behavior is then controlled by or shaped by external stimuli only. No counterweight to environmental influences is available and the person is at the mercy, then, of the environment.

PERSONALITY DISPOSITIONS

Findings from the fields of personality and temperament theory support the idea of these general orientations toward the stream. Singer (1984) factor analyzed "large numbers" of questionnaires on private experience. One factor was labeled "positive constructive orientation to inner experience"; a second was labeled "guilty-dysphoric"; and a third factor reflected "distractibility" or poor attentional control. In a recent thorough review on the biology of personality, Zuckerman (1994) posited three major personality dimensions based closely on Eysenck's (1994) work. These are labeled (1) psychotocism-impulsivity-unsocialized sensation seeking, (2) neuroticism-emotional and, (3) extraversion-socialbility. Berzonsky (1990) has synthesized the massive literature on personality styles into three different types: An information orientation is marked by active efforts to seek information relevant to the self before developing firm personal beliefs or commitments. A normative orientation, on the other hand, reflects reliance on local peer group standards when developing personal beliefs. These individuals cordon off core aspects of their identities to protect them from potential disconfirmation from incoming information. A diffuse orientation relies on immediate situational and contextual demands, rather than being directed by internal principles and information. From the point of view of regulating the relationship to the stream of thought, the first dimension would be composed of people who vacillate between approach and avoidance reactions to the stream (diffuse orientation). The second group are those well disposed toward the stream (the information-oriented individuals), and the third group—the extroverts—are the normative types who prefer an instrumental and avoidant stance toward the stream. These personality types may also be related to attitudes toward risk when making decisions under conditions of uncertainty. There are the so-called risk-neutral (diffuse) individuals, the risk-averse (normative), and risk-seeking (information) types.

Given the well-documented links between personality types and health, it might be the case that health status, too, is influenced by one's orientation to the stream. Singer and Bonnano (1990), for example, argue that introverts or sensitizers were more prone to cardiovascular disease, while extroverted, repressor types were more prone to general psychosomatic and neoplastic disease.

I do not wish to claim that personality is reducible to one's orientation to the stream of thought. I merely wish to suggest that that orientation colors or influences personality style. One walks a tightrope or a knife's edge in navigating one's relation to the stream. One must be open to it but not submerged in it. One must maintain some distance from it, but not ignore it completely. Presumably, the ideal relationship involves a little of both of these stances.

NEUROLOGY OF SELF-REGULATION

These broad personality dispositions are supported by neuropsychological systems implicated in mediating self-regulatory strategies. Gray (1991) distinguishes two limbic centered emotional systems in the regulation of behavior: (1) the "behavioral inhibition system," involving septo-hippocampal structures and (2) an approach system centered on the medial forebrain bundle. This latter system mediates impulsivity, while the BIS mediates anxiety. When too much information is the problem, anxiety levels rise, triggering the BIS, which then inhibits ongoing processing. When arousal and information levels are too low, the approach system is triggered and the BIS is itself inhibited.

Gray's model of self-regulation is consistent with Pribram and McGuinness's (1975) model, discussed in Chapter 4, as well as other models (see below), which postulate approach-avoidance systems in the regulation of behavior. These two systems are in mutual inhibitory balance and are the basis for personal autonomy and choice or attention. The approach system is biased toward selection from the external or public realm of the stream, while the withdrawal system is biased to select from internal, private portions of the stream. These biases do not preclude overall biases concerning the stream (public and private) in general. An overall approach bias toward the stream will result in an introverted profile, while an overall avoidance bias toward the stream will result in an extroverted profile.

Sanides (1969), using evidence gathered from cortical architectonic studies, suggested that the six layered isocortex evolved from two primary sources of undifferentiated cortex: the archicortex of the hippocampus and the paleocortex of the olfactory lobe. Goldberg (1987) applied these findings to the problem of the evolution of the frontal lobes and postulated two premotor frontal systems. The lateral system controls action that is environmentally based and therefore data driven. It functions in a responsive mode rather than in an prospective or predictive mode. The medial system is used to control action when memory can be used to allow prospective control of behavior. The medial system, in other words, supports voluntary intentional behavior. Goldberg's arguments are consistent with Sanides's original formulations concerning the whole brain.

Sanides recognized sets of cortical-growth rings, with regions of more recent evolutionary appearance surrounding the older zones. He made the reasonable assumption that areas with more differentiated cortex (e.g., greater number of layers, greater number of cell types, greater connectivity, etc.) were the more recently evolved cortex. Comparative and more recent studies confirmed these reasonable assumptions (Pandya & Barnes, 1987). These progressively more differentiated architectonic fields represent waves of neuronal differentiation originating from the two cores. Patterns of connectivity and intracortical laminar distribution of projections are used to try to infer function. The medial paralimbic trend originates in the hippocampus, proceeds through the anterior cingulate gyrus, the inferior temporal lobe, the supplementary motor area in the paralimbic cortex, and then becomes prefrontal cortex. Interestingly, outputs of the basal ganglia appear to be directed toward this medial system, indicating a role for the medial system in support of voluntary behaviors. The lateral parainsular trend begins in the olfactory cortex and proceeds through the insula, the superior aspect of the temporal lobe, the second somatosensory area, and then becomes the parietal lobes.

Goldberg suggests that the lateral trend may function to respond to and register external sensory information, while the medial trend may specialize in analyzing interoceptive and mnemonic information. In humans, the lateral trend is much more highly developed in the parietal lobe, while the medial trend predominates in the frontal lobe. These two systems are considered to be in mutual inhibitory balance in an approach-avoidance equilibrium. Personal autonomy depends on maintaining some kind of oppositional pole to the parietally centered external focus. That oppositional pole is centered in anterior networks and, at the cognitive level, depends on memory abilities.

Given the anatomical structures associated with the lateral trend (e.g., the olfactory system, the insula, somatosensory association area, superior temporal sites, and the parietal lobes), it would seem that the system is specialized for sensory and perceptual processing and attention to external stimuli as well as some language functions. The medial trend, on the other hand, appears to be specialized for memory (hippocampus), emotional processing (limbic), and higher-order functions like planning, initiation, and voluntary control of behavior (supplementary motor area and dorsolateral prefrontal). One consistent functional theme associated with these medial structures is that all are crucial for "personhood." If lateral-system structures were impaired, we would say that the person had lost his ability to process stimuli or to understand language adequately, but we wouldn't feel that the personality had been centrally affected. With lesions to medial structures, on the other hand, personality "functions" (memory, emotion, volition, planning, etc.) are directly affected. Indeed, profound personality disturbances often occur with lesions to the frontal lobes.

CLINICAL SYNDROMES

Evidence from the neurological clinic supports these broad inner versus outer dichotomies in human functioning and self-regulation. The classical form of anosognosia involves patients who are apparently unaware of their illness. Sometimes this unawareness may take the form of outright denial of hemiplegia. Anosognosia usually involves lesions in the parietal lobes. This implies disinhibition of the frontal lobes, so patients attend only to internal stimuli and ignore or are unaware of the "external" hemiplegia. In the related syndrome of unilateral neglect the associated lesion is usually in the parietal lobes and usually on the right side. Patients fail to attend to any stimuli on the left side of space. They may not eat anything on the left side of the plate or they may tie only their right shoe or fail to address someone on the left side of space relative to their position. Since the anterior and the posterior systems are theorized to be in mutual inhibitory balance, we assume that a lesion to the posterior system would disinhibit operations of the anterior system, so patients overrely on the anterior system when performing cognitive operations. Since the anterior system is biased toward internal components of the stream, patients are focused only on internal stimuli and neglect external stimuli. Since the lesion is one-sided, the neglect is asymmetric also. Lhermitte (1986) described an interesting syndrome associated with bilateral frontal damage, which he called the "environmental dependency syndrome." Patients with bilateral frontal damage often display a remarkable tendency to imitate the actions of those around them (echopraxia). This involuntary "imitation behavior" may extend to the most embarrassing gestures such as urinating in public or chewing paper. Imitation behavior may also be associated with "utilization behavior." If the examiner, for example, places a set of everyday objects in front of patient with instructions to not pick them up or use them, the patients nonetheless do just that! If one of the objects was a comb, patients would involuntarily pick it up and begin combing their hair. Utilization behavior may even extend to dangerous objects such as hypodermic needles. Here patients might even attempt to give themselves injections. It is as if the patient's behaviors were controlled by salient objects (objects that capture the attention) or salient behaviors in the environment and not by the patient's own decisions. The patient's behaviors, in other words, were environmentally driven rather than personally chosen. The patient's very autonomy is threatened. Since the lesion is in the anterior system, the posterior system is disinhibited and the patient, therefore, is biased toward the pickup of external stimuli. That is the patient uses external stimuli to regulate behavior.

In attention deficit disorder, children find it difficult to self-regulate because frontal lobe development lags behind parietal-lobe development. Thus, children find themselves at the mercy of a posterior system operating without anterior inhibition. Whatever is salient captures these children's attention. They attempt to find ways to inhibit the operation of the posterior system so they may find some sort of equilibrium. Occasionally, children with these kinds of attentional disorders benefit from catecholaminergic medications like Ritalin. These medications may work by activating inhibitory circuits in the frontal lobes. When frontal functions are activated, they can oppose the countervailing influences of the externally tuned posterior systems and thus confer a degree of behavioral autonomy on the child.

Evidence for bilateral-frontal dysfunction has been documented in several other disorders (e.g., advanced stages of Parkinson's disease, obsessive-compulsive disorder, and Tourette's syndrome) associated with disinhibition of posterior systems. Obsessive-compulsive disorder (OCD) is an illness characterized by high levels of anxiety, intrusive and disturbing thoughts, and elaborate rituals (e.g. hand washing or repeatedly checking the stove before leaving the house). OCD is also very frequently associated with Tourette's syndrome. Sacks (1992) described the remarkable case of a surgeon with Tourette's syndrome who displayed a whole series of behaviors that seemed to be compulsive in nature and environmentally driven. This man felt compelled to repeatedly touch or react to virtually any salient event in his environment, including a lampshade, walls, people, the windshield in his car (even while driving), and the windshield in his Cessna airplane (even while flying!). He would often, in an attempt to retain some independence, "mark his territory" off from the environment around him by drawing (with his foot) a circle around his person. When the demands of the environment became too great he would resort to extraordinary measures to oppose those demands. For example, Sacks describes a bizarre scene during a case conference between the surgeon and a group of his colleagues. While listening to the comments his colleagues were making on the case, the surgeon curled up on the floor, kicked his foot into the air and then tapped his colleagues on the shoulder with his foot! Such extraordinary behaviors begin to make some sense when we see them as strategies adopted by the patient to oppose the "pull" of external or environmental contingencies. By opposing these powerful environmental forces (and perhaps innate approach tendencies) the anterior system (in cooperation with the posterior system) helps us to maintain some degree of personal autonomy.

While patients with Parkinson's disease seem less susceptible to environmental demands than patients with Tourette's syndrome, they nonetheless still exhibit some remarkable instances of environmental dependency. They may, for example, be unable to walk forward without the examiner placing some salient stimulus in the path of the patient (e.g., an imaginary line on the floor). Presumably, the "line" invented by the examiner becomes a salient stimulus, which then can guide the patient's subsequent progress. The effects of saliency on these patient's behaviors can often be quite striking. I once tested a series of PD patients who were nondemented but who, nevertheless, performed poorly on tests of frontal function. If I indicated to a patient that testing would take place in "that office at the end of this long hall," the patient would begin to literally bound down the hallway toward the indicated site. I had to hold down many patients to keep them from falling while rushing toward the office at the end of the hall. If, however, I said testing would take place "right around the corner here," patients would shuffle peaceably around the corner toward the indicated site. When the idea "end of the hallway" was made salient, that image/idea seemed to control the patient's ambulation and directional goal and the patient was under a compulsion to achieve that goal immediately.

Maintaining a balance between activation levels in the anterior and the posterior systems clearly functions to protect personal autonomy and assist in self-regulation. It protects us from an overemphasis on either interior events alone or external, perceptual stimuli only. It allows for selective processing of relevant stimuli and the screening out of irrelevant or harmful stimuli. Thus, people appear to possess neuropsychologically based regulatory systems that allow them to optimally interact with the mass of variable mental objects in the stream of thought.

MULTILEVEL SELECTIONISM

Given the evidence for two major neuropsychological systems that reflect two strategies of dealing with the variability inherent in the stream of consciousness, do these systems participate in attentional or selectional processing? Internal selection, for example, may be handled by the anterior or medial system. Each level of the system—from hippocampus to cingulate and supplementary motor area and frontal cortex—participates in selection. Each anatomical area specializes in processing certain types of information and therefore preferentially selects out a corresponding portion of the stream. Internal selection may proceed by inhibition of the posterior system which then releases the tonic inhibition on the anatomical elements of the anterior system. The release from inhibition allows for generation of mental images associated with each anatomical area. For example, the hippocampus handles memory processing, the limbic system handles emotional information, the anterior cingulate regulates limbic sites, the inferior temporal lobe mediates higher aspects of visual processing, the supplementary motor area supports initiation and volitional behavior and the frontal lobes are important for personality, planning, insight, and so on. The lateral system seems to be more involved in sensory processing—the olfactory lobe in smell, the insula in somatosensory and pain processing, the second somatosensory area in touch and kinesthesia, and the parietal lobes in body schema, somesthesis, and higher aspects of spatial cognition. Both the insula and the superior temporal sites are also involved in language comprehension. When the medial system is activated the lateral system is inhibited and processing of external information is suppressed. Conversely when the lateral system is activated, medial structures are suppressed and processing of internal information (memories, affects, imaginative reverie etc.) are suppressed.

In the last three chapters we have sketched out James's concepts concerning a selectionist account of memory and personal identity. The Mind, says James, is at every level a theater of simultaneous possibilities. The stream of thought always presents us with an array of possible states or selves. We select out from this array a series of selves that arrange themselves in a kind of functional hierarchy, which includes a dominant self that handles daily interactions and a series of subordinate selves that reflect various drives, goals, purposes, and roles that may or may not be in conflict with the dominant self. The dominant self must compose this array of internal selves and develop a system of self-regulation vis-à-vis competing selves. How does the dominant personality achieve any kind of self-regulation of this internal hierarchy of selves? How does it remain at the top of the hierarchy? Where and how do competition and selection between possible selves take place? Some amount of selectionist processing in memories and "selves" may occur during the dream state (dreaming will be discussed in Chapter 10), but the process of recollection itself must surely involve creative interactions between self and memory.

CHAPTER 9

Phenomenology of Recollection

RECOLLECTION AND SELF

Recall that there are good reasons to believe that a selectionist account will provide a new understanding of the relation between memory and identity. First, selectionist theory implies that the organism will come equipped with rich internal structure. Since selection requires a rich array of traits that persist through time, it necessarily requires a rich internal structure in the organism possessing those traits. Those traits, in fact, constitute the structure of the organism. In an organism with some amount of self-awareness, that rich internal structure must be counterposed to the less differentiated external structure. The individual is, in effect, defined by the boundary between internal and external structure. Blind variation and selective retention result in the gradual emergence of structure, complexity, and individuals. The emergence from selectionist processes of complex, bounded, self-organizing entities supports the idea that selection promotes individuated entities. How does selection promote a sense of human individuality and personhood—an individual with definite boundaries and a unique constellation of traits? Once again, memory is crucial. Personal consciousness is not possible without an autobiographical memory. Autobiographical memory constructs a personal consciousness via recollection.

In the simplest sense, recollection is a collecting or gathering up of disparate images or memories from the stream of thought into a single entity.

It is selection at the level of the self or the appropriation to the self of a subset of a vast array of images from outside the self. According to Bergson's selectionist theory of remembering, the emergence of a sense of internal structure and identity occurs during the process of recollection. Recollection requires an act of will, which involves an active suppression of the intense attachment to the external environment and the present: "To call up the past in the form of an image, we must be able to withdraw ourselves from the action of the moment, we must have the power to value the useless, we must have the will to dream" (1888/1988: 83).

During an act of recollection, there is an initial relaxation of inhibition, which makes available to the rememberer a vast array of images experienced as different from the rememberer, since the images are believed to be images from the past. This relaxation of inhibition is associated with a letting-go of the intense attachment to the external world and a refocusing on things internal. Next, a secondary process of inhibition is triggered, which corresponds to the selection process. This selection process, in turn, constitutes "retrieval" or the preservation of a subset of those "past" images into the domain of the present or the self. Thus, recollection is a selection of images and an incorporation of those images into the structure of the self. While the memory images themselves come out of a nonspatial realm, they nevertheless enter the spatial realm when they adhere to an identity or self-concept.

In what follows I provide a phenomenological analysis of this recollective process or experience in order to explore its effects and the ways in which it allows for the incorporation of selected images or new structures into the structure of the self. I am interested in how the recollective process "feels" and what its effects are on the rememberer. The incorporation of images into the self structure is crucial for development of an identity or self. Casey (1987) specifies the meaning of recollection in the following way: "Collection derives from the Latin collecta, a "gathering together" and still more primordially, from colligere, literally a "binding together" (as is signified in the English verb to colligate); whereas re signifies back or again. In a primary act of recollection, I bind things together, keep them in a gathered unity so that I can return to them again and again (1987: 292–93). For Plato, recollection involved access to and selection from the eternal ideas, and then an identification of self with those ideas. Recollection was a remembrance of knowledge once grasped but lost through the forgetfulness of the unenlightened life. Plato's doctrine concerning recollection was very selectionist in spirit, as it asserted that what we call learning was really a remembering and that our knowledge comes by contact with the eternal ideas. Remembering, in the Platonic view, is therefore a spiritual faculty par excellence since it delivers to us the eternal element in the human soul. For Socrates and Plato, recollective experience was direct experience of the providential Self who knew firsthand, so to speak, about the eternal ideas. Recollection put the individual in touch with an infinity of images, that is, "all things."

> *The soul, then, as being immortal and having been born again many times, and having seen all things that exist, whether in this world or in the world below, has knowledge of them all; and it is no wonder that she should be able to call to remembrance all that she ever knew about virtue, and about everything; for as all nature is akin, and the soul has learned all things there is no difficulty in a man eliciting out of a single recollection all the rest—the process generally called learning—if he is strenuous and does not faint; for all learning and inquiry is but recollection.* (Edman, 1928: Meno, 81c–82a)

To recollect is to take a step back from the world. Now, since the ego must relinquish its attachments, it grieves its losses and mourns its defeat by an agency more powerful than it. It is interesting that the Indo-European root for memory is *(s)mer*—to mourn. The Germanic root is *smerd*—pain. The Latin is *memor*—to be mindful. The Sanskrit root is *mendh*—to direct the Mind. The Greek root is *merimma*—care or sorrow. Memory's deep roots extend back into mourning.

If recollection does not lead one to "all things"—a vast array of images from which to select—then the selection process does not proceed and no new structures are incorporated into the organism. According to Plato, most of the men of this world have lost their memories and therefore they suffer:

> *But all souls do not easily recall the things of the other world; they may have seen them for a short time only, or they may have been unfortunate in their earthly lot, and, having had their hearts turned to unrighteousness through some corrupting influence, they may have lost the memory of the holy things which once they saw. Few only retain an adequate remembrance of them; and they, when they behold here any image of that other world, are rapt in amazement; but they are*

> *ignorant of what that rapture means, because they do not clearly perceive. For there is no radiance in our earthly copies of justice or temperance of those other things which are precious to souls; they are seen through a glass dimly; and they are few who, going to the images, behold in them the realities, and these only with difficulty. . .* (Edman, 1928: Phaedrus, 249c)

The fourth century Platonic philosopher Boethius preserved this view of memory for the Christian West when he argued that maladies of the soul always involved the forgetting of one's divine origins. Recollective experience frees us from undue dependence on the external world. It temporarily severs the intense attachment to the external environment or world. Without the ability to periodically withdraw from the world and its attachments, the external environment would pretty much constitute the primary determinant of human behavior—it would excessively control our behavioral options. Recollective experience allows us to avoid losing ourselves in the world. It does so by a process Casey called "in-gathering" (Casey, 1987). I call this process related to in-gathering, incorporation. James called it appropriation. It is like what Bergson termed translation—the transformation of nonspatial memory images into a spatially delimited self. If recollection proceeds to a state of detachment from the incessant flow of the stream, then recollection bestows interior peace. What is occurring during these varied recollective states is the coming to presence of an autonomous self.

There are three important processes related to recollections' support of incorporation. One process concerns our most primitive sense of place. Place is where presencing or incorporation happens. A second process linked to the phenomenon of presencing and the sense of place is the experience of personal boundaries. Boundaries may be an effect of the gathering-in process, which occurs during recollection. Boundaries and the sense of place are undoubtedly linked. The third process related to recollection's incorporation powers is the emergence of a sense of intimacy, self , or identity. Intimacy implies communion between two individuals and so recollection ultimately leads one out of the self and into genuine community.

SELF AND PLACE

Place for Aristotle comes first in the order of things because it exerts an active force on the things it situates. That is, place has the power to influence things, to change them, or even to bring them into being. According to Casey (1987), Aristotle was the first to suggest that the localization of things or objects is due to an inherent power of place itself, that place itself structures these things or objects. Heidegger called the coming into being of things and of a self "presencing." Presencing, in turn, depends on place. We can provisionally define place in memory, then, as that being or portion of the stream that calls forth or supports the presencing of a self and situates that self.

While presencing/incorporation depends on place for its initiation, the entire process of presencing of a self seems to depend on recollection. Place initiates a process of recollection or anamnesis. Anamnesis, in turn, puts the individual in touch with the eternal forms of "all things." The encounter with this infinity of images or possible worlds supports selection of structures from the vast array and then an incorporation of those structures so that the presencing of a self occurs in a bounded place.

Place is that power that evokes a presencing of a self—it calls a self into being. We experience place's influence in our lives as an interior experience of recollection. Thus place is an essentially private experience. Place triggers recollection. Integral to place's recollective and presencing powers is its regionalization and boundary-making powers. A boundary occurs where presencing begins. At the point where regionalization occurs is where human beings are compelled to build. Building is an instinctual response to presencing within a set of boundaries. With the establishment of boundaries comes a sense of interiority and intimacy—a sense of an inside (the boundaries) and an outside (of the boundaries). Interiority establishes an infinite dimension of depth within these boundaries. From a selectionist point of view, the bringing into presence of a self characterized by intimacy, depth, and interiority must coincide with selection's ability to accumulate beneficial variations relative to a given ecological niche. As mentioned above, the accumulation of beneficial variations constitutes a lineage or history of the species and the species' abilities to incorporate salient features of the local environment. Each organism, therefore, carries within itself a "replica" of the local environment as well as the collective memory of its species. Yet this collective memory is placed in service to the needs of the individual organism. The "memory" of each person, therefore, supports the paradoxical experience of infinite depth and personal intimacy.

BOUNDARIES

A process linked to the phenomenon of presencing and the sense of place is the experience of personal boundaries each one of us carries about our persons. Boundaries and place's regionalization powers are undoubtedly linked. According to Heidegger, a boundary is an effect of presencing. A building sometimes instantiates these boundaries in stone—it allows us to see our presencing, rather than just to experience it unconsciously.

Hartmann (1991) has argued that in the realm of mental activity, thickness or thinness of boundaries relates to the degree of compartmentalization and distinctness (thickness) in cognitive processing versus a preference for merging and blending (thinness) of contents and processes in mental activity. In the realm of personality types, boundaries would refer to the thickness or thinness of the barriers people erect between themselves and the other portions of the stream. People with thin boundaries would be more open to the stream as well as other influences/people, while those with thick boundaries would be less open to the stream/influences. In either case, boundaries define where a person begins and ends relative to the ongoing stream.

James spoke of remembering as a process which the self infused with a special intimacy, inwardness, and warmth. These are the salient qualities of the subjective experience of remembering. With every act of recollection, I must revert to myself alone as I will only find the images there and nowhere else. The past presents itself to me as ordered for me, about me, and so on. Nevertheless, because inwardness, subjectivity, and solitude are precisely the experiences that we all share, the subjective experience of remembering leads one back into community. This time the community is a community of selves, each with a unique solitude and a unique history. At the heart, then, of the phenomena of presencing lies an intimacy that discloses a single unique and solitary individual. The intimacy is at once definitive of this particular unique individual, but it is also the only authentic experience common to all. Intimacy implies communion with another. The original other is in fact a portion of the stream called place (since place called the self into presence) and the ground of intimacy and, therefore, community is simply the encounter between place and presence.

In summary, remembering from the selectionist point of view involves selection of a subset of—past—images from a vast array of images that emerge into mental life when the Mind relinquishes its intense attachment to things external to the rememberer. Another term for this selectionist remembering is recollection. A phenomenological analysis of recollection reanalyzed recollection as an incorporation of selected images into the structure of the self. This incorporation process probably coincides with the emergence of boundaries that define the place of the self—where it ends and where it begins. They also promote a feeling of immense depth and "intimacy." All of these phenomenological properties of recollection add up to something we call a personal identity.

MYTH, RECOLLECTION, AND IDENTITY

Interestingly, mythological accounts of recollection ascribe profound spiritual and psychological effects to the experience. It is worth reviewing some of the salient themes in this literature as these accounts are congruent with the claim that memory and recollection are crucial for personal consciousness and autonomy.

As mentioned above, Plato believes that learning or the acquisition of knowledge is really recollection. In the *Meno* and elsewhere, Plato argues that the soul before it comes into this world has access to the Pure or Eternal Ideas. But when the soul is reincarnated, it drinks from the spring of Lethe and forgets the knowledge it had obtained from contemplation of the Ideas. "Lethe" is generally assumed to refer to the river or spring of Lethe, of forgetting. The Greek word and its cognates meant originally "to hide oneself" or "being hidden." Lethe the Goddess was, according to Hesiod, a daughter of Eris the Goddess of dissension, chaos, and discord. All who are born into mortality must first drink from the river Lethe so that they forget everything about their pre-earthly existence. A deceased soul would descend to Hades to find a white cypress tree. To the left of the tree is the spring of Lethe. The soul is advised by the Greek Book of the Dead to avoid even moving close to that spring.

The spring to the right of the tree is the cold water spring of Mnemosyne—Goddess of memory—who stands next to it in an eternally open gesture to those who thirst. The spring is shown on a classical vase that depicts the underworld as flowing through a mask. The dead person at that point will experience unbearable thirst and the goddess will invite the soul to drink "from my eternally flowing waters" to the right of the tree. When the soul drinks from the well of memory, it can then show itself in the underworld as a spirit kindred to the gods themselves.

Mnemosyne, the goddess of memory, was the source of every recollection, no matter how mundane or profound. She was the sister of Kronos and Okeaonos and mother (spouse to Zeus himself) to the Muses—those goddesses of all of the arts and sciences. Mnemosyne, in other words, was understood to be the ultimate source of all of what we would now call creativity and the creative arts. Something of the awe with which Memory was held in the ancient world is reflected in the works of the great Roman orator and philosopher Cicero. In the *Tusculan Disputations*, he describes memory as that faculty from which all discoveries and inventions flow and he then goes on to say that he is not inquiring into cases of spectacular or exceptional memories. The inventive and mnemonic powers of the most ordinary human being are a wonder. "I am speaking of the average memory of man, and chiefly of those who are engaged in some higher branch of study and art, whose mental capacity it is hard to estimate, so much do they remember" (quoted in Yates, 1966: 44). For the ancients, invention was a function of memory—not a separate power in itself. Just as the Muses issued from Mnemosyne, so do the arts come from the generative powers of memory.

Mnemosyne was also considered to be one of the great primordial goddesses—one of the six or seven Titans from whom all the other gods and goddesses were descended. Hesiod referred to Mnemosyne as the comforter to the hurt and the afflicted. Mnemosyne was omniscient and imparted to her devotees a very specialized form of knowledge, namely the knowledge of origins, of beginnings and geneologies. It is this knowledge, the knowledge of origins and "becomings" that allows one to rival the gods in creativity. As Eliade put it: ". . . recollection does not seek to situate events in a temporal frame but to reach the depths of being, to discover the original, the primordial reality from which the cosmos issued" (1963: 120). Recollection delivers to the individual the ability to pass over to another dimension—the past—and to return from it bearing knowledge and treasure.

The knowledge of origins always came in the form of images. When images are experienced, manipulated, expanded, amplified, combined, inhibited, and so on, new ideas, new entities, new knowledges emerge. "Thinking back" always results in the production of a prodigious number of images. Images for the ancients were forms, marks, or simulacra (forma, notae, simulacra) "corporeal similitudes," icons, and so on. Aristotle asserted that the soul thought in images. Plato thought the process of recollection included recovery of images that referred to the Eternal Ideas. Augustine found the spacious palaces of memory to be populated by an plenitude of images. Out of this generative matrix of imagery sprang the Muses or, in a word, creativity. In any case recollection was considered the sacred method by which one could achieve godlike creativity and illumination.

This central Greek myth concerning the powers of recollection and the fate of the soul passed into the Platonic and esoteric intellectual traditions in the West and in to Islam and Judaism. Boethius salvaged the doctrine of recollection in his "Consolation of Philosophy" during the transitional period of the collapse of paganism and the rise of medieval Christianity. The doctrine of recollection became known as the practice of *dhihkr* (remembrance or recollection) in the mystical and Sufi traditions of Islam. Medieval Jewish scholars incorporated the doctrine of recollection into the mystical philosophy of the Kabbalah. In the Platonic tradition, recollection had the ability to heal the suffering self because it revealed the self's true, divine nature. While loss of memory is the soul's true malady, recollection's ability to lead the soul back to its true nature or homeland constituted a healing journey for the soul.

Recollection allows the soul to return to God—the stillpoint of the turning world—and thus allows the soul to avoid becoming disturbed by the ups and down of the wheel of fortune. Memory sees all, records all, witnesses all.

> *That which remembers, or that which is aware of all things, must be a principle always present to (anubhu) all things, and therefore unaffected by the duration in which these events succeed one another. We are thus reduced to a Providence (prajna) or Providential self or spirit (prajnatam) as the ultimate source on which all memory draws, and with which whoever attains to the same uninterrupted omniscience must be identified as in Prasna Upanishad IV.10.* (Coomaraswamy, 1977: 58)

Eliade has called "The Hymn of the Pearl," found in the Acts of Thomas, the central Gnostic myth and it is a myth concerning recollection. It depicts the effects of recollection on a lost soul. In the myth, a prince comes to Egypt from the East seeking "the one pearl, which is in the midst of the sea around the loud breathing serpent." In Egypt, he forgot his princely origins and his quest for the pearl and became enslaved to local authorities. But his parents and friends heard of his plight and sent a reminder to him that he was a son of kings and not a slave and that he was in search of the pearl. The reminder jogged his memory and he set out to liberate himself. He recovered the pearl from the serpent and returned home. On his return, he puts back on the "luminous garment" he had taken off when he first left on his journey.

Memory in "The Hymn of the Pearl" seems to function as a kind of liberating force for the self. When the prince remembered that he was the son of royalty, he quit his slavery and retrieved the pearl. Recollection leads one to discover one's true identity or true self. The discovery of one's true self through recollection is also a prominent theme in the religious and mythical traditions of India. According to Eliade (1963: 115), many of the Hindu myths relating to memory can be summarized in the following terms: "1. A spiritual master falls in love with a queen or is made prisoner by women. 2. In either case there is a physical love that immediately provokes a state of amnesia in the master. 3. His disciple seeks him out and through a series of symbols (dance movements, secret signs, mysterious language) helps him to recover his memory, that is consciousness of his identity. 4. The master's forgetfulness is assimilated to death, and vice versa—his awakening, or anamnesis, proves to be a prerequisite for immortality" (1963:115). As with the "The Hymn of the Pearl," forgetting is equivalent to sleep, loss of self, loss of immortality, blindness, and the fall into existence. Remembering, on the other hand, frees the soul into immortality.

One familiar myth from the Hindu tradition is known as the "Roar of Remembering" or the "Roar of Awakening":

A pregnant tiger, ravenous with hunger, prepares to pounce on a herd of sheep. She miscalculates, pounces, and the sheep scatter. She then goes into labor but dies during the birth. The baby tiger is left motherless. When the sheep return, they find the baby tiger. Taking pity on the baby, they decide to feed it and raise it as one of their own. The baby tiger is brought up as a sheep. It learns the sheep language ("bahhh . . . bahhh"), it learns to eat and digest grass, and it learns the ways of the herd. It experiences some developmental difficulties, however. It finds it difficult to deny its attraction to meat. It underestimates its own strength and so frequently hurts its sheepish playmates. It cannot totally reconcile itself to the ways of the herd, preferring to go it alone much of the time. Worst of all, it does not experience the same responses and feelings the rest of the herd experiences when attacked by its enemies—the predators. In particular, it feels no fear and looks for a fight.

Despite these developmental difficulties, the baby tiger grows into adulthood under the watchful eye of the herd, completely convinced that it was, in fact, a sheep. Then one day, when the baby tiger had become a young adult, the herd was attacked by another tiger—a female tiger his own age. All the sheep scattered as usual and he was left alone to face the attacking tiger. When she saw him, she grew angry and contemptuous and demanded: "What are you doing here amongst all these sheep?" He answered: "Bahhh . . . bahhh . . ." and then absentmindedly nibbled some grass. The female tiger then flew into a rage, grabbed him by the scruff of the neck and dragged him for miles until the day passed into night. She brought him to a pond. The sky was filled with the light of a full moon. "Look at your reflection in the pond and tell me what you see," the female demanded. The tiger was worried because he did not understand why he so easily understood tiger language. He looked into the pond and fainted.

When he woke up, he found himself in a cave filled with the carcasses of dead animals and piles of meat. He felt sickened by the sight but could not leave. He felt strangely at home. The female tiger then demanded that he eat. "Bring me some grass," he said. The female tiger then forced open his mouth and stuffed some raw, bloody meat down his throat. As he sat there gagging on the meat, he felt the blood from the meat slowly oozing down into his gullet and deep from within his bowels he felt a trembling stirring. As it gathered force, it developed into a roar of joy which came pouring out of his being. He had remembered his tigerhood.

Like the other Hindu myths concerned with memory, remembering is afforded the unique power of revealing to the rememberer his true nature and identity and that knowledge brings with it profound joy. The theme is very much like the theme enunciated in the Gnostic "The Hymn of the Pearl."

The notion that memory is linked with liberating knowledge for the self is found in the Norse tradition also. In the Norse traditions, the Well of Memory is found in Mimir, the land of the giants. The land of the giants represents an intermediate realm between the land of the Gods and the land of Men. Mimir exacts a price from all those wishing to drink from the Well of memory. Odin, like Christ on the cross, was bound to the World Tree (Yggdrasil) and paid Mimir's price (Odin's right eye) in order to drink from the Well. After drinking from the Well he discovered runes or the art of writing as well as other spiritual powers.

CONCLUSION

Whether one follows my phenomenological account of recollection or the above-reviewed mythological account of the powers of recollection, some common themes emerge. Recollection introduces the subject to a vast realm of potent "images." The subject selects a subset of these, which then pass into the service or the structure of the self. In the mythological accounts recollection is a spiritual experience that allows subjects to remember their true divine nature, while in the phenomenological account, recollection involves a turning inward and the assimilation of new structures into the self. While the selectionist paradigm cannot be said to "explain" these accounts of recollection, at least they are not inconsistent with selection theory. They point, furthermore, to the deep and indissoluble links between memory, freedom, and self.

CHAPTER 10

Dreaming

Another "place" where the selection of memory images takes place is in the dream. According to Foulkes (1985), dreams are involuntary symbolic acts that share some characteristics of other cognitive systems. Like Fodor's input modules (1983), dreaming is automatized and hard wired—or supported by specialized anatomical circuits—and it outputs a specialized information product, namely the dream. But dreaming cannot, like other modules, be considered to be informationally encapsulated. Other modules have access to the workings of the dreaming module. Nevertheless, dreams are organized mental products produced by a specialized neuropsychological system. Given these facts concerning the neuropsychological system that produces the dream, it is likely that dreaming itself performs some specialized function for the cognitive system. What is that function?

Dreaming involves an operation or computation on memory images to produce a structured product called the dream. If memory images are processed in the dream, and if the processing is selectionist with cumulative retention of certain memory images, then dream mentation, like waking mentation, must be producing information content or knowledge for the dreamer. Examining that content could give us a clue as to function.

DREAMING AND MEMORY

Many investigators into the dream state have remarked on a special relation between memory and dreaming. While most of these investigators have argued that dreaming involves the consolidation of memories in support of learning, some have argued that rapid eye movement or REM sleep, the physiological state associated with vivid dreaming, involves a form of reverse learning—the elimination of useless memory images. In either case, dreaming is conceived as a form of memory processing. Recent content analyses of dreams confirm Freud's early observation that some of the dream's imagery is composed of recently acquired memory fragments.

Foulkes (1985) believed that the source of most of the dream's imagery came from memory stores, since incoming sensory information was blocked during the dream state. Foulkes went on to argue, however, that the memory images that entered the dream were "decontextualized" memory fragments—that is, memories of happenings without the contextual information concerning time and place. Hobson (1988) noted that one of the striking characteristics of the dream state is the marked increase in access to memory; the dream might actually be said to be hypermnesic. This hypermnesia while in the dream is contrasted with the equally dramatic amnesia for most dreams upon awakening. Winson (1985) pointed out that REM sleep in most other mammals is associated with a characteristic theta rhythm in the hippocampus. The hippocampus is known to be crucial for formation of new memories in most animals. Winson argued that dreaming represents a form of "off-line" or unconscious memory processing. Basic behavioral strategies, important for survival, are acquired during a critical period in early development. REM sleep, according to Winson, somehow constructs or rehearses these basic strategies of behaviors. On the face of it, therefore, there seems to be a relation between memory and the dream, though the nature of the relation is unclear.

What does the dream do with these memory images? Freud's work on the so-called dreamwork and dream censorship suggests that the memory images are used in the service of self-deception. One of the co-authors of the often cited "activation-synthesis" hypothesis on dream formation (Hobson, 1988) asserts that dreams, contrary to what Freud argued, have no disguised meaning at all. Rather, Hobson argues that dreams represent an attempt on the part of the cortex to "synthesize" or to interpret random activation patterns generated in subcortical centers. Dream content reflects these synthesizing efforts of the forebrain; thus, dream content directly represents the dreamer's drives, fears, and associations. For Crick and Mitchison (1986), dreams are meaningless. REM sleep functions in order to reduce fantasy and the interaction of related ideas and concepts in memory. REM sleep, loosely speaking, is a form of reverse learning. The cognitive system is purged of elements that might interfere with efficient functioning. It therefore might even be harmful to remember dreams. Winson's ideas concerning the dream were mentioned above and are consistent with Freud's notion that dreaming dredges up disguised and unacceptable material from the unconscious. Winson's concept of the unconscious, however, is not so concerned with unacceptable material as is Freud's version of the unconscious. Winson places the emphasis on cognitive and memory processing and has little to say about how the dream handles unacceptable information.

Despite these largely negative assessments of Freud's work on memory images in dreams, there is still some acceptance of the view that dreams have some kind of regulatory role to play in the emotional economy of the self (see Moffitt, Kramer, & Hoffman, 1993; Hartmann, 1996). Hartmann, for example, carefully documented the role dreaming played in recovering from traumatic life events. The dream's ability to symbolize or contextualize the overwhelming emotions associated with the trauma helped the dreamer to integrate the trauma and then let it go. This integrating capacity of the dream may be related to a basic operation of the dream: the making of connections. If the fundamental nature of dreaming involves a process of making connections in memory more broadly and more disparately than in the waking state (Hartmann, 1996), then dreaming may be a powerful tool for the rewriting of old scripts or unacceptable autobiographical memories. Hartmann's hypothesis furthermore implies that some people will be better at this (i.e., dreaming and memory rewriting) than others.

Bergson (1901/1920) suggested that the dream involved a lifting of the normal inhibitory stance of the brain, a relaxation of the usual "interested" and outward-oriented stance of the individual. Once restraints are relaxed, ". . . these repressed memories, feeling that I have set aside the obstacle, raised the trap-door which held them back below the floor of consciousness, begin to stir. They rise and spread abroad and perform in the night of the unconscious a wild phantasmagoric dance. They rush together to the door which has been left ajar. They all want to get through. . ." (1901/1920: 117).

The stirring up of previously "repressed" memories that flood the cognitive system corresponds to the proliferation phase in the Bergsonian cycle of remembering. In the case of sleep, the actual number of images that flood the system must be greater than the number generated during the proliferation phase in waking life, since proliferation in waking life is constrained by a single cue or perceptual image. That is, images must relate precisely to the current perceptual image or cue that initiated the cycle in the first place. This is not the case with sleep, since sleep involves a relaxation of outward-oriented attention; thus, images can be related or matched to one another more loosely. In the waking state, proliferation helps to contextualize the cue and thus helps to locate the target image. In the dream state it is not clear what the cues are. They may be affects or fleeting internal sensations. Thus, the contextualization of these "cues" is complex. The generation of associates to an affect may produce a more complex array of images than the generation of associates to a perceptually derived cue.

After proliferation comes selection. We have seen that, in the case of waking perception, the images that are selected are those that match in some way a feature in the environment—or, more precisely, the ones selected are those that are useful for the free action of the individual. In the case of dreaming how are images selected? There is now no outward orientation, no interested stance, no vivid perceptual sensations, and no sustained attentional effort with which to guide the selection process.

Because of the sensory blockade associated with REM sleep, internal and affective sensations dominate the dream state. Thus, selection is guided by these affects. The dream is a peculiar combination of a set of fleeting external impressions (which occasionally escape the sensory blockade) along with another set of internal (affective and physical) impressions that combine with related memories into a distinct mental experience. Memories that best match (or are in harmony with) the general affective state of the organism are more likely to be selected (or combined with the sensations) than other memories.

The mechanism of dreaming, then, is identical to the mechanisms of remembering and of waking perception. They are all selectionist processes. In each case, a selection is made from a vast array of memory images and the selected images are added to the cumulative store. The cumulative store encodes knowledge about the world and the dreamer. That is why dreams can be called "credible world analogs" (Foulkes, 1985). They embody (through memory) knowledge about the world of the dreamer. In the case of perception and of remembering, those memories are selected that most exactly match the cue and are useful in some way to the organism. In the case of dreaming, selection is governed mainly by the general affective state of the individual. By "general affective state," Bergson presumably meant to point to whatever emotions and feelings are activated during sleep. Thus, the cumulative store would encode the dreamer's emotional intelligence. We know today that limbic system structures are preferentially activated during rapid eye movement sleep (Gillin, Buchsbaum, & Wu, 1993), so Bergson's emphasis on affects in relation to dreaming seems well placed.

Bergson is at pains to point out that it would not be accurate to state that dreaming was like perception minus the ability to reason. According to Bergson (and this differentiates him from most other theorists of the dream, including the psychoanalytic theorists), the higher cognitive faculties are not suspended during dreaming. The dreamer unwisely attempts to impose some kind of logical sense on the vast array of images that the dream unleashes. Bergson's explanation for the bizarre in dreams is not unlike that of modern investigators (Hobson, 1988; Foulkes, 1985) who suggest that the dream ego or (a left hemisphere–based) interpreter tries to make sense out of a jumble of random images and so constructs the narrative of the dream—such as it is. But Freud and many modern students of the dream assign more importance to the narrative than I believe Bergson would have done. Since images, in the Bergson theory, are closer to the affective state of the organism than the narrative is (they are proliferating in response to the affective state), images contain more information about the affective state than does the narrative. The interpreter uses the images to construct a story. The story may or may not reflect the true affective state of the organism. It depends on how closely the images match the state of the organism.

In summary, according to Bergson, the dream represents the sum total of normal mental life minus the effort of attention. When the brain relaxes its inhibitory powers with the onset of sleep, the cognitive system is flooded with a vast array of memory images. The memory images proliferate in relation to any fleeting external impressions that might be perceived as well as any internal affects or sensations. As the proliferation of memory images proceeds, some undergo the process of selection: they are matched to whatever sensations/affects are salient at the moment. The rest are inhibited. The matches are not exact, due to the lack of attentional effort. An interpreter tries to arrange the proliferating memory images into some kind of narrative. Occasionally, one of these narratives survive, into waking life and that is what we call a dream. The crucial point for deriving a theory of the function of the dream is that the criterion of selection involves affective rather than perceptual cues. We conclude that the dream specializes in processing affective memories.

Now affective memories are important only for a particular personal consciousness. The function of the dream must therefore involve the processing of information relevant to the concerns of an individual consciousness. Since dreaming, furthermore, involves the selection and cumulative retention of autobiographical memories, the function of the dream must concern the construction and selection of possible identities or selves.

SELECTION OF "SELVES" AND REM SLEEP

The biology and phenomenology of REM are consistent with the hypothesis that processing occurs around construction and elimination of characters or selves. In the mature state, dreams are predominantly visual phenomena. Reports of smell, taste, or pain sensations are very rare (Hobson, 1988; Symons, 1993). Affective levels are usually higher relative to normal waking life. Dream images usually involve characters or persons interacting in fairly mundane ways (Hall & van de Castle, 1966; Van de Castle, 1994). Many, though not all, dreams have a story like or narrative quality. The dreamer (the dominant self) is engaged in some struggle to do something and faces threats or obstacles along the way (see Foulkes, 1985; Hobson, 1988; Symons, 1993, for discussions of the phenomenological properties of dreams). Many of these phenomenological properties of the dream appear to involve the processing of autobiographical memories. The fact that dreaming is hypermnesic, egocentric, and emotional suggests that some kind of processing around self-related affects and memories is happening.

REM sleep is reliably associated with vivid dreaming. It is a cyclic phenomenon (occurring approximately every ninety minutes during sleep), with specially dedicated brain mechanisms that regulate its appearance and suppression (Hobson, 1988). REM is characterized by sensory blockade, by a complete absence of muscle tone in the antigravity muscles, and by brain-wave activity like that in the waking state (McCarley, 1989). Although incoming sensory information is blocked, many regions of the brain (especially the limbic and orbitofrontal regions) sustain very high brain-activation levels (Gillin, Buchsbaum, & Wu, 1993). McCarley (1989) has suggested that the phasic bursts of rapid eye movements are associated with pontine-geniculate-occipital (PGO) waves. These are spikes that were originally associated with sites in the brainstem pons, the lateral geniculate nucleus of the thalamus, and the occipital cortex—all visual centers. Steriade (1989), however, has pointed out that PGO waves are recorded in most thalamic nuclei and therefore probably serve to activate much wider areas of the cortex than the occipital system. REM onset is controlled by cholinergic systems, and the REM state itself is associated with a characteristic electrical rhythm in the hippocampus called the theta rhythm—though such a rhythm has not yet been definitively identified in humans (Winson, 1985). The reader may have noticed that all of the brain sites activated in REM are also sites in Sanides's "medial trend"(see Chapter 7). We found that the medial trend supports functions normally important for personality and identity. The medial trend involves the hippocampus, limbic and anterior cingulate sites, inferior temporal sites, supplementary motor areas, and frontal cortex. Medial trend specialties include memory, affect, volition, higher-order visual imagery, and other personality-related functions—all prominent phenomena in dreams. The lateral-trend mediates sensory and language processing—two functions that are blocked or often absent in REM sleep. The inhibition of lateral trend functions implies the activation of medial-trend functions. Thus, dreaming may be primarily a medial-trend function. Since dreaming, in any case, involves the activation of medial-trend sites, personality support functions must be involved in the cognitive processing that occurs in REM.

Dreaming, then, can be described as an arena where the selection of memory images proceeds in service to identity selection. Selection, of course, implies selective retention of some images over others. I suggest that autobiographical memories are the ones in question. Given the fact that affective brain sites are active during REM sleep, while most incoming sensory information is blocked, selection takes place on the basis of whatever affects are present/activated. The selection process may be usefully described or detailed via Freud's dreamwork: Condensation involves the reduction of a mass of imagery into one set of images. Displacement involves the transfer of libido or emphasis from one set of images to another—that is, one set of images are highlighted at the expense of others. Repression is inhibition of nonselected images. Selection of memory images in the dream is in service to a given self, which makes up the dream ego. Whatever images are not consistent with this ego's aims are eliminated or revised; thus, a dominant self emerges.

If the foregoing arguments are correct, then dreaming involves a kind of play or experimentation with alternate identities. Inhibitory processes are relaxed, while a wealth of affectively tinged memories become available to the dreamer. These are obviously fine conditions for an exploration of alternate identities. But Bergson argues that the conditions of the dream allow us to explore more than simple or everyday alternative identities. According to Bergson, dream-memories do not reside in the brain or the head of the dreamer. Rather, the dreamer and all the rest of us are always steeped in a sea of memory-images. We become aware of this sea of imagery when, as in the dream, the brain relaxes its inhibitory grip on our consciousness. The dream, in other words, allows us to explore portions of that sea of imagery (or James's stream of thought) we are not usually aware of.

DREAMING AND PSI ABILITIES

It is an interesting fact that both Bergson and James, the two thinkers who were most responsible for introducing selectionist thinking into models of Mind at the dawn of the twentieth century, were both seriously interested in psychical research. Both also linked psi phenomena to dreams. Organized scientific inquiry into parapsychological phenomena, such as telepathy, clairvoyance, and the like, began in the 1870s in England, under the leadership of the moral philosopher Henry Sidgwick, a man much admired by James. Sidgwick was energetically assisted by F. W. H. Myers, who subsequently devoted a great deal of time and effort to the study of psi abilities. He, too, seemed to think of the brain as a filtering or selectionist device of some kind. Bergson was elected president of the British Society for Psychical Research in 1913. His presidential address was titled "Phantasms of the Living and Psychical Research," echoing the title of one of Myers's books. Except for the presidential address and some paragraphs in his last work, *Two Sources of Morality and Religion* (1932), Bergson, wrote very little about parapsychological research. We will therefore have to rely on his philosophy to infer why he believed in the reality of psi phenomena.

Like Bergson, James became involved with the British Society for Psychical Research in the early 1880s. He was also a founding member of the American Society in 1884. His involvement in actual research on psychic phenomena seems to have been more extensive than that of Bergson. From the 1880s on he participated in numerous séances, sittings, experiments on telepathy, clairvoyance, automatic writing, and so on. According to Gerald Myers, a recent biographer of James, James linked the possibility of psi with the idea of multiple personality or multiple selves within one individual, since this condition suggested that consciousness or identity could transcend the body (Myers, 1986). If personal identity depends on memory or a recollecting consciousness, as Bergson, Locke, and others surmised, and if one brain or body—one individual that is—could be the site of several different selves, then personal identity may not depend on sameness of substance (i.e., one body, one soul). James, therefore, deduced that identity could not be reduced to brain. "It is clear already that the margins and outskirts of what we take to be our personality extend into unknown regions. Cures and organic effects, such as blisters, produced by hypnotic suggestion show this as regards our bodily processes; while the utterances of mediums and automatic writers . . . though usually flimsy and incoherent . . . as the present writer believes, occasionally show a knowledge of facts not possessed by the primary person" (quoted in Myers, 1986: 376).

Bergson, too, argued that consciousness and brain could not be identified with one another. Consciousness overflowed the brain on all sides, according to Bergson. Memory ensured that consciousness extended decades back in time, while desire and imagination (both partially dependent on memory) extended forward in time. The brain merely acted as an inhibitory device regulating the volume of mental imagery consciousness had to deal with. He went on to suggest that since consciousness could not be reduced to the brain, it might therefore survive the demise of the brain. James and Bergson concluded that one's own identity may not be identifiable exclusively in terms of one's own body but may extend into experiences (or streams of thought) associated with others—whether these others are hidden selves within one's own subconscious or those associated with other bodies altogether. Such cases would then be considered telepathy.

Although it is difficult to piece together the theories James and Bergson held regarding the possibility of psi, it appears that they believed that their selectionist stances concerning Mind and consciousness were compatible with theories of anomalous phenomena like psi.

If the brain is simply an inhibitory or filtering device (i.e., the organ of attention to life), as James and Bergson argued, then relaxation of that inhibition (as in such abnormal states or hypnosis or dreams) would promote access to the sea of "images" present in the collective stream of consciousness. If, as we have argued in this book, selection in the realm of memory holds true, then people develop cumulative memory stores from which variant memories are generated in each selection cycle. According to this theory, election of target variants constitutes consciousness. Because people in a given culture possess similar cumulative memory stores, the variants they generate in a given memory cycle will also contain similarities, and when these variants match one another, those persons experiencing the match will experience one another's thoughts. The extreme case would occur when two people transiently select relatively similar or even identical "selves." Such an event is theoretically possible, given the selectionist assumptions we've outlined in the previous pages. The most likely place where two people would generate similar associates or select similar selves is in response to an affective cue, since the emotions are largely universal and stereotypical. As we have just seen, selection in the realm of the dream is guided by the affective state of the dreamer. Thus, if we find, for example, reports of psi phenomena associated with dreaming, then Bergson and James's theoretical speculations would be supported.

Van de Castle (1994) recently reviewed the literature on psi and dreams. He noted that several prominent dream experts, from Freud and Jung to Stekel and Boss, have all strongly insisted on the reality of paranormal dreams. Louisa Rhine collected over 7,000 accounts of ESP experiences from people who responded to appeals published in various media. The majority of the accounts involved dreams. Numerous other anecdotal accounts of paranormal dreams have been published so that there is no question that dreaming, if the accounts can be trusted, seems to be particularly well suited for the production of psi experiences.

Van de Castle's review of the Maimonides project on paranormal dreams is instructive since he participated in these as a subject. In this set of studies, an agent attempted to telepathically send a target stimulus to a percipient who was asleep in a locked, soundproof bedroom. The percipient's sleep stages were monitored with the usual EEG equipment. The target stimulus was randomly selected from among hundreds of colored art reproductions of paintings or sculptures. The same target picture was used for the entire night. The agent (or sender) was in a separate bedroom, which was at one point in this series of experiments located fourteen miles away from the percipient's bedroom. No contact was allowed between recipient and sender once the experiment began. When an experimenter (who was monitoring the EEG) noted that the percipient had entered a REM sleep stage, he signaled (via intercom) the sender who would then concentrate on the target stimulus and attempt to send it to the recipient. After ten or fifteen minutes of REM activity, the monitor woke the percipient (via intercom), who then reported whatever dream content he had been experiencing. The dream report was tape recorded and later transcribed for analysis. In the morning, the percipient would be shown a collection of eight art reproductions, one of which had been the target stimulus. The percipient rank-ordered each picture in terms of its likelihood that it was the target stimulus. Finally, the percipient assigned confidence ratings for each picture in terms of how well the pictures matched the content or emotions of his dreams. An outside panel of judges were given dream transcripts and the collection of art pictures and rated them for correspondences or matches in content. There were thirteen major studies carried out in the Maimonides project and statistically significant results were obtained in nine of those studies. Child (1985) has also reviewed statistical and experimental design issues connected with these studies, and found them satisfactory (see also Bem & Honorton, 1994).

Although these results are suggestive, they are certainly not conclusive, and the existence of psi effects is still quite controversial. Bergson and James apparently believed that the effects were real and that they were connected with the diurnal change in selection criteria from hard-headed daytime pragmatic formats to dreamtime affective reflection. Given the robust effect sizes, these modern investigations into psi effects associated with the dream state should probably be continued.

Although dreams may involve the selection of memory images and possible selves, selective processing of selves or identities is also constrained by extrapersonal group and social factors. We consider some of these cultural constraints in the next chapter.

CHAPTER 11

Selection, Self, and Culture

SELECTION AND CULTURE

Human culture appears to exhibit all of the standard selectionist diagnostic criteria: Cultural artifacts vary in their properties. As a consequence of these properties, some are preserved and reproduced while others are not (selective retention). Those preserved are placed within a tradition that represents the cumulative wisdom surrounding those particular artifacts. A given cultural tradition or domain evidences inner-directed change, development, and evolution of forms—historicity, in a word. Thus, on the face of it, cultural practices may obey selectionist principles. Certainly, several authors have suggested as much (Rindos, 1985; Schmid & Wuketits, 1987). Of course, an equally strong case can be made for the view that cultural change depends on instructivist methods of information acquisition. Classical, instruction-based learning, in other words, occurs in the cultural realm. Certainly, imitative forms of learning occur all the time in cultural settings. I, however, will avoid the controversy over cultural learning and approach the issue from the bottom up, asking what the implications for culture might be if memory itself obeys selectionist principles.

Most people would agree that, in a fundamental way, learning and culture depend on memory. But culture's dependence on memory tells us nothing special about culture, since virtually every other human practice depends on memory. It might be argued, however, that culture depends on a particular kind of memory. If correct, that claim would be more interesting because it would place the critical link between Mind and culture in a specific, neuropsychologically identifiable system. One of the most interesting proposals along these lines has been Dawkins's (1976) elucidation of the meme.

MEMES AND SEMANTIC MEMORY SYSTEMS

Memes were originally defined by Dawkins as the basic Darwinian units of cultural transmission that obeyed classic selectionist principles. "Examples of memes are tunes, ideas, catch phrases, clothes fashions, ways of making pots or of building arches" (1976: 206). Dennett later pointed out that these newly defined replicators were "complex ideas that form themselves into distinct memorable units—such as the ideas of arch, wheel, wearing clothes, vendetta, right triangle, alphabet, calendar, the Odyssey, calculus, chess, perspective drawing, evolution by natural selection, impressionism, 'Greensleeves,' deconstructionism" (1995: 344). It is interesting to note that both Dawkins and Dennett limited their examples of memes to very restricted classes of memories, namely semantic memories and procedural memories. People's semantic and procedural memory systems carry the public or cultural knowledge of a society. If we conceive of memes as information complexes that jump (perhaps via imitation or through communications media) from Mind to Mind, then meme complexes compete for access to the memory systems of individual human beings. Meme systems evolve due to differential replication rates exhibited by competing meme complexes. Blackmore (1996) has pointed out that there are far more memes in the world than there are vehicles to carry them. No meme complex can replicate without a vehicle to support transmission. Thus, competition around access to these vehicles is fierce. That competition is what we usually call the struggle or competition of ideas. What are the criteria that enable meme complexes to win out over other meme complexes when competing for access to appropriate vehicles for replication and transmission?

Those memes that win out in the competition will, on average, fit criteria that are tightly constrained by memory-based factors. Dawkins (1976), for example, suggested that copying fidelity, longevity, and fecundity determine a meme's survival value. Other authors have suggested such "traits" as learnability, assimilability, ease of communication, tendency to be transmitted and other such characteristics (see Heylighen, 1995). I will comment only on Dawkins's criteria, as these are the most general.

Copying Fidelity

In order to enhance copying fidelity, memes must engage in various mnemonic strategies that increase their chances of infecting the host. Many of these strategies have been studied as encoding processes in the memory literature. Among these are recruiting additional attentional resources during the encoding process, structuring to-be-remembered study material and utilization. Meme complexes may also structure themselves in such a way as to capitalize on available attentional resources or to create optimal encoding conditions. Some meme complexes may, for example, multiply the number of semantic connections that can be made with other (existing) semantic networks. Memes that promote repetition or rehearsal in the host enhance the saliency of the meme complex during the encoding process. Schematizing or abstracting the content of the meme complex provides for better copying ability as well as ease of communication. A meme complex must be able to recruit attentional resources either via the embedding of salient cues within the complex itself or by the encouragement of elaborative processing procedures once encoding begins. It must be capable of making rich semantic connections with existing meme networks (assimilability), and it must be abstract enough to allow for compatibility with the most common network architectures.

Once these initial conditions are met, the meme complex will either win or lose in the competition for entry into the host's cognitive systems. Its "aim" is to be encoded in such a way as to promote long-term retention in the host.

Longevity

The longer a meme resides in a suitable host or vehicle, the greater its chances of (a) restructuring the host's Mind so that incompatible memes will be excluded and (b) infecting other hosts or vehicles. These maneuvers clearly depend on the ability of the meme complex to resist the assimilation of competitor meme-complexes into the host's memory systems. Memory capacity might also be increased by elimination of existing memories—a kind of internal editing process that is guided by the purposes of the dominant meme complex. Crick and Mitchison (1986) suggested that dreaming and REM sleep might involve the systematic elimination of parasitic memories. If they are right, memes might compete for access to the machinery of REM sleep on the "hope" that this would enable them to restructure the memory system for their own purposes.

The ability to resist intrusions by competitor memes into the host's memory system probably depends primarily on rejection of belief systems that are "incompatible" with the host's current belief system. People, of course, differ in their openness to foreign or incompatible belief systems. Much of what is known as group consciousness centers around the need to regulate access to Minds. To the extent, therefore, that a meme complex can insulate itself from public or rational evaluation, it will remain impervious to change. Self-monitoring in this sense is the enemy of memetic longevity.

Fecundity

As meme complexes increase in number and variety, so do their chances for survival. For any given meme, the greater the number of copies available, the greater its chances for survival. Clearly, cultural forms such as communications media, the visual arts, print and broadcast media, the Internet, and even fax machines all determine propagation rates of copies of memes.

Although it appears that memes must compete for access to semantic memory networks, it may be that their prospects for longevity and reproducibility would be better served by assimilation into autobiographical memory systems. As seen in previous chapters, the self-concept (at least the dominant self) is densely interconnected with virtually every other cognitive and physiological system. If a meme could infect the self structure, its control of the host would be greatly enhanced and so would its prospects for reproducibility and longevity. Thus, from the meme's point of view, the critical link between Mind and culture is not only semantic memory systems. Autobiographical memory systems are critical as well because these systems give rise to the self.

We thus arrive at the conclusion that the real influence of memory on culture comes from the autobiographical memory system. Memory's impact on culture is probably mediated through its intermediate impact on personality. Unfortunately, as William James argued, selectionist autobiographical memory mechanisms give rise not to a single self but to a hierarchy of selves. While not a problem for meme-assimilation strategies (memes, presumably, would initially target the dominant self), this hierarchy of selves is potentially troublesome for overall self-regulation. According to selectionist theory, new possible selves are constantly being generated, and existing selves compete for dominance in the hierarchy. This competition, if not well-regulated, could create crippling internal conflict. It may be that various cultural forms function to assist the individual in coping with the problem of psychic or emotional self-regulation.

Many cultural forms function, in part, as a kind of escape valve. They accommodate or channel the excess productions of the human Mind. In particular, some cultural forms and practices may also function, in part, to mitigate potential breakdowns in personality structure. Here we leave the realm of semantic memory systems and the memes that structure them and begin to focus on larger cultural and social arrangements.

CONSTRAINTS ON "POSSIBLE SELVES"

One way to assist the individual in self-regulation is to put a constraint on the number of possible selves that person could entertain in the internal hierarchy. Another way would be to regulate the interactions between the selves in the hierarchy. The division of labor (or specialization), for example, serves, in part, to constrain the number and range of social roles or social selves available to a person in a given culture. Specialization accomplishes this regulating function by creating slots or roles within which people must fit. A modern society, for example, needs auto mechanics, while agrarian societies need farmers. Mumford (1966) saw the roots of specialization and technology in the prodigious symbolic capacities of the Mind. This excess of symbolic energies spill over into other channels we call cultural: religion, mythology, dance, ritual, the arts, and, of course, tools and other technologies. This formulation reverses the vulgar Marxian notion that cultural forms are determined by the technological level of a society. In Mumford's formulation, the excess symbolic capacity of the Mind leads to technological innovations. Mumford points to the example of agriculture: The "intense neolithic subjective concentration on sexuality in all its manifestations, expressed first in religious myth and ritual, still abundantly visible in cult objects and symbolic art" led to the neolithic pre-occupation with the natural processes of birth, growth and death, seasonal cycles, plant and animal cycles etc. "Plant selection, hybridization, fertilization, manuring, seeding, castration were the products of an imaginative cultivation of sexuality, whose first evidence one finds tens of thousands of years earlier in the emphatically sexual carvings of Paleolithic woman: the so-called Venus's" (1966: 81). Cultural forms or practices—technical specialization, in particular—have the side effect of helping to channel or constrain the excess symbolic capacities of the Mind. Among those excesses is the overproduction of identities or possible selves. At bottom, all the strategies available to a person for dealing with the instable, overproductive human psyche are forms of self regulation. Cultural forms may also assist individual human beings in this task of self-regulation.

We saw in earlier chapters that humans use selection to self-regulate their interactions with what James called the stream of thought—that great and endless sea of mental variability. We as selves deal with the excess variability in the stream of thought in two general ways: approach or withdrawal—openness of closedness. One cannot simply ignore or close oneself off from the contents of the stream—else no innovation, development, or change would be possible. Similarly, one cannot maintain a constant openness to the stream, else one would simply be deluged and then psychically destroyed by the materials and energies of the stream. Thus, we have evolved a kind of strategy wherein we approach the stream for new infusions of energy and then, once we have had our fill, we attempt to dam up the waters until we need to approach them once again. This self-regulation strategy is evident in the daily attentional fluctuations we all experience. As we saw in previous chapters, information processing involves an initial orientation to novelty and a proliferation in working memory of an array of novel mental representations, and then selection from this array. Selection may be accomplished via repeated sampling for similar stimulus attributes. This selection for redundancy involves an inhibitory stance toward the rest of the stimulus array; that is, a closed-off stance toward the rest of the contents of the stream. In summary, we can adopt either an open stance toward the stream or a closed stance. Optimal self-regulation probably involves a balance between the two stances.

Both of these strategies lead to interesting consequences for cultural life as well. Nietzsche, for example, described two tendencies operating in ancient Greek culture, which profoundly influenced all the arts.

Nietzsche's categories included the Dyonisian, after the Greek god, Dyonisius, of orgiastic excess; and the Apollonian after the god Apollo, the god of distance, order, and form. The Dyonisian experience would, in our terms, imply openness to the contents of the stream. The Apollonian experience, on the other hand, would imply a more closed or distant and abstract disposition to the stream. While the ancient Greeks recognized these extreme tendencies at the heart of the human soul, they nevertheless sought to emphasize balance, proportion, and justice. Greek culture helped average Greek citizens find a balance between open and closed moralities, as they strove to achieve autonomy and freedom.

In any case, these general reactions to or orientation toward the stream can also be traced in the evolution of social and cultural systems. Once again, Bergson was the first to develop this approach to the evolution of human societies. Like Mumford, he saw developments in culture as reactions to the excess intelligences of the Mind. As in the case of individual mental economy, the all-important question is how to profitably relate oneself to the vast and incessant imagery produced by the Mind—that is, what to do with the excess. Similarly, with the case of social economy and cohesion, the pressing question is how to deal with the outpouring of mental productions from huge numbers of people endowed with human levels of intelligence and consciousness.

OPEN AND CLOSED SOCIETIES

Bergson's late work *The Two Sources of Morality and Religion* (1932) contains his reflections on social economy and development, and the related problems of political and moral philosophy. He contrasted two forms of social and cultural orders. The closed society relies on an ethic of obedience to establish cohesion and order, while the open society looks first to dialogue in negotiating conflict. In our terms, the closed society would have the individual self-regulate by enforcing an ethic of obedience among internal and competing motivations and selves, while the open society would foster dialogue between competing selves or tendencies in the individual.

The Closed Society and Static Religion

According to Bergson, insect social systems are regulated entirely by instinct, whereas human social systems, given the added element of individual choice and intelligence, must be regulated by something beyond instinct but which nonetheless supports the social instinct.

Human society, "fresh from the hands of nature," as Bergson often puts it, is necessarily a "closed society" in both its ethics and its religion, since it is primarily concerned only with survival, order, stability, and defense against other communities. It is turned inward on itself, as it expects only aggression and war from other societies. Its social rules are intended to maintain order in society and to defend against outsiders. The closed society is oriented toward maintaining stability and is intransigently conservative. It relies on "obligation" rather than raw social instinct to attain a degree of social cohesion and to enforce obedience to its laws. One might say obligation is the expression of the social instinct among humans—though this would not be entirely correct in Bergson's view.

The "totality of obligation" represents the moral and cultural code elemental human society relies on to function. Obligation is based on social pressure, custom, and habit—not reason. Underlying obligation is the raw social instinct that reached its highest development not in humans but in the social insects. Humans rely less on instinct and more on intelligence than do the other animals. This has important consequences, as we will see, for the development of religion and social experience.

Bergson argues that social life evolved along two trajectories (insects and higher vertebrates) because both had specializations in the use of implements or tools. For the insects, the implements were organic extensions of their own biologies, while for humans the tools were products of intelligence. Tools imply specialization and division of labor within a species and specialization, in turn, gives rise to social roles and social organization. Evolutionary development of implements among humans is dynamic, propelling further social change—even in closed societies. In the insects, however, implement utilization is governed by instinctual forces and is therefore essentially immutable.

Intelligence in human societies plays a consistently disruptive role. It serves to further stimulate an already overstimulated human psyche. It leads to technical innovations that promote changes in the division of labor and so disrupts basic social roles and divisions. It vastly complicates the task of self-regulation. It leads to a thoroughgoing egoism by showing individuals that they are worthy and have dignity, regardless of their social station. Intelligence is like a can of worms for the closed society—the society based on obedience and obligation—because it makes individuals aware of their own mortality and frailty. Intelligence allows individuals to conceive unusual ideas and to plan for the future, but it also allows them to see that the best-laid plans often come to naught. Intelligence, in a word, deepens awareness of all things, including suffering. Its connection to its object, technical implements, furthermore, leads to constant innovation in the division of labor and, therefore, social roles. The fluidity in social roles creates a similar fluidity in the Jamesian hierarchy of selves. In order to maintain social order and stability, the closed society needs to counter the disruptive effects of intelligence and it does so by producing what Bergson called static religion.

Static religion counters the deep individualistic impulse created by intelligence with a panoply of powerful deities that forbid self-centered behavior and command conformity to social obligation (the lists of do's and don'ts characteristic of closed societies). The subordinate selves in the Jamesian hierarchy are kept in line by enforcing obedience to one dominant self. To the disruptive awareness of death, static religion counters with an assertion of life after death; to the disruptive influence of frustration of plans and hopes, static religion opposes the image of a powerful and compassionate deity who can bring good fortune. In sum, static religion is a defensive reaction against the socially disruptive or dissolvent effects of excessive intelligence. It allows for an effective attachment of the individual to life. It supports the project of self-regulation by imposing a relatively rigid creed of obedience and sanctioning the paradigm of a single dominant self who lords it over a number of subordinate selves. It is, in fact, the highest moral expression of the closed society.

Bergson is not asserting that religion evolved simply to treat the pain of existence. Static religion did indeed evolve to treat the pain of existence, but it also developed to handle the socially disruptive effects of a dynamic and unruly intelligence. Static religion channels the excess products of intelligence into cultural forms that cancel out, at least temporarily, the disruptive effects of the Mind's prodigality. Here Bergson is providing an analysis of human intelligence and culture as a generative capacity that is constantly evolving or proliferating into new and compartmentalized functions (e.g., the myth function), and which is rooted in the fundamental necessities of biological and social life. Intelligence appears here as a set of rapidly evolving capacities that call forth responses and new faculties to deal with the superfluity of impulses emanating from within intelligence itself. This intelligence, therefore, can even devise ways to fool or deceive itself. The original business of intelligence was utilitarian, but its dynamic nature constantly undermines the ethics of obligation and obedience. Intelligence requires a new ethic, which supports the tasks of self-regulation and social cohesion without the negative consequences of using an ethic based on obedience only. At this point, Bergson introduces his explication of the open society.

Even though closed societies are conservative and hostile to outside influence, these societies nevertheless evolve. The gods of static religion, for example, change and grow as a function of the circumstances closed societies face. The constant development of tools and the associated instability in social roles fuels the change that occurs in closed societies. Within closed societies, there are also individuals who find a way to develop beyond the strictures of the closed society. For Bergson, this last source of change is crucial for the moral and spiritual health of humanity.

The Open Society and Dynamic Religion

Who are the individuals that break through the ethic of obligation and subservience to the gods of static religion? For Bergson, they are the mystics. The mystics point the way to the open society. The open society is an ideal, never yet fully realized. Such a society, even when it is contained within the bosom of a larger closed society, can be a locus of moral and religious creativity. Cohesion in the open society is achieved through political negotiation and conflict resolution. Negotiated cohesion (as opposed to imposed cohesion) introduces new and dynamic social forces into the system and therefore disrupts institutions based on moral closure and an exclusionary ethic. Open societies require and promote religious experience that is individualistic, open, and joyful. Dynamic religion consolidates this individualistic vision, which is open and embraces all of humanity as kin.

Dynamic religion represents a different way of channeling the excess energies of the psyche than does static religion. Static religion uses the myth-making function to heal psychological wounds inflicted by intelligence. The result is a kind of social cohesion and stability. Dynamic religion achieves a new equilibrium with the stream by recapturing the creative impulse that brought humanity into existence in the first place. Closed society organizes itself against further infusions of the life giving energies of the stream of consciousness, also known as experiential time or the life force or, simply, reality. The open society, on the other hand, allows for the translation of these transpersonal energies into the consciousness of a single individual. The open society opens itself up to the stream of images that constantly press in on the human psyche. It has found a way to allow them in without being destroyed by them. Multiplication of diverse modes of being in the open society allows these excess energies to flow into the culture without permanently disabling it. Diversity in the social realm mirrors, to some extent, diversity in the psychic realm of the stream. Such a project of openness to the terrifying energies of the stream requires great psychological strength and profound nobility of spirit, but it has been pursued at all times and by all peoples. "Suppose we say that it embraces all humanity; we should not be going too far, we should hardly be going far enough since its love may extend to animals, plants, to all nature . . ." (Bergson, 1932: 38).

Bergson is careful to show that such a complete morality, an all-embracing love, cannot be achieved through rational or personal exertions alone. For Bergson, these mystics open themselves up to reality, the élan vital, the life force in all its profligacy, creativity, and prodigality. What they experience and what they communicate are a great emotion and deep vitality. By communicating that emotion to the surrounding culture in a closed society, mystics prod that society into growth and response. The great emotion experienced by mystics is crystallized by those surrounding them, into a new ethical doctrine, but couched in the language of static religion, since those are the representations available to these people. Over time, these new insights are lost and static religion ossifies into dead ritual.

How does humanity evolve into this open morality and to dynamic religion? Bergson said that if a few follow the great mystics and if others imagine they would do likewise on occasion, then this already means a great deal. But he warned that it was not by preaching love of neighbor that the closed circle would be broken. It was not by simply expanding our narrower feelings that we could embrace all of humanity and nature. "The truth is that heroism may be the only way to love. Now heroism cannot be preached, it has only to show itself and its mere presence may stir others to action. For heroism itself is a return to movement and emanates from an emotion—infectious like all emotions—akin to the creative act. . ." (1932: 53).

Such are the general outlines of Bergson's application of selectionist principles to the social dialectic. On the one hand, we have all around us systems of closed moralities—a panoply of static religions all conducing to docility, obedience, and conformity. These are the very values that prevent a sustained challenge to business as usual. Obedience-based ethics, however, cannot work for individuals endowed with human-level intelligences. They cannot assist us in the project of achieving workable self-regulation. Societies based on obedience and obligation, therefore, are inefficient. On the other hand, developing within these very systems of closed morality we find individuals who cultivate a different ethic, based on dialogue and negotiated solutions to conflict. When practiced as a strategy of self-regulation, this alternative ethic leads to individuals who are not crippled by serious internal conflict and who can thus focus their energies more effectively on external problems.

The great mystics—the Buddha, Jesus, Mohammed, Gandhi—were great leaders, people of action. They were not just world-denying contemplatives. Their legacies last, they influence whole cultures for centuries. Mystics pull us into the struggle for social justice—not out of anger or hatred, but out of love and compassion. Their general tenor is quiet, enduring joy, the communication of joy, liberation and peace. A sympathy for all things animate and inanimate. A tremendous energy, ability, and determination to organize and make a difference. An orientation toward inspiration from life or nature. A willingness to become a vessel, an instrument of "God's grace." An aversion to violence without losing the capacity for violence. An active service to the higher self in every person encountered. These are the virtues we see surfacing over and over again in the testimonies of the mystics. They confronted the contents of the great stream of consciousness and brought back treasure.

References

Anderson, M. C., and Bjork, R. A. 1994. Mechanisms of inhibition in long term memory: A new taxonomy. In *Inhibitory processes in attention, memory and language*, ed. D. Dagenbach and T. H. Carr, pp. 265–326. San Diego: Academic Press.

Augustine. 1979. *Confessions.* Loeb Classical Library First Edition, 1912. Cambridge, MA: Harvard University Press.

Baddeley, A. 1986. *Working memory.* New York: Oxford University Press.

Baldwin, J. M. 1889. On selective thinking. *Psychological Review* 5 (1):1–24.

Barkley, R. A. 1997. Behavioral inhibition, sustained attention, and executive functions: Constructing a unifying theory of ADHD. *Psychological Bulletin* 121 (1): 65–94.

Barkow, J. H.; Cosmides, L.; and Tooby, J. 1992. *The adapted mind: Evolutionary psychology and the generation of culture.* Oxford, UK: Oxford University Press.

Bem, D. J., and Honorton, C. 1994. Does Psi exist? Replicable evidence for an anomalous process of information transfer. *Psychological Bulletin* 115 (1): 4–18.

Benson, D. F., and Blumer, D. 1975. *Psychiatric aspects of neurological disease.* New York: Grune and Stratton.

Bergson, H. 1888/1988. *Matter and memory.* N. M. Paul and W. S. Palmer, trans. New York: Zone Books.

——. 1901/1920. Dreams. In *Mind Energy.* H. Wildon Carr, trans., pp. 104–134. Westport, CT: Greenwood Press.

——. 1932. *The two sources of morality and religion*. R. A. Audra and C. H. Brereton, trans. London: Macmillan.

Berzonsky, M. D. 1990. Self-construction over the life span: A process perspective on identity formation. In *Advances in personal construct psychology*, ed. G. J. Neimeyer and R. A. Neimeyer, pp. 155–186. Greenwich, CT: JAI Press.

Bjorklund, D. F., and Harnishfeger, K. 1995. The evolution of inhibition mechanisms and their role in human cognition and behavior. In *Interference and inhibition in cognition*, ed. F. N Dempster and C. J. Brainerd, pp. 141–173. New York: Academic Press.

Blackmore, S. 1996. Memes, minds and selves. *Journal of memetics: Evolutionary models of information transmission*. www. cpm. mmu. ac. uk/jom-emit. html.

Blumer, D., and Benson, D. F. 1975. Personality changes with frontal and temporal lobe lesions. In *Psychiatric aspects of neurologic disease*, ed. D. F. Benson and D. Blumer. New York: Grune and Stratton.

Brewer, W. F., and Pani, J. R. 1983. The structure of human memory. *The Psychology of Learning and Motivation* 17:1–35.

Broadbent, D. A. 1958. *Perception and communication*. London: Pergamon Press.

Brown, R., and Kulik, J. 1977. Flashbulb memories. *Cognition* 5: 73–99.

Brownowski, J. 1977. Human and animal languages. In *A sense of the future*, ed. J. Bronowski, pp. 104–131. Cambridge, MA: MIT Press.

Bruner, J. 1994. The "remembered" self. In *The remembering self: Construction and accuracy in the self narrative*, ed. U. Neisser and R. Fivush, pp. 41–54. Cambridge, U.K.: Cambridge University Press.

Byrne, R., and Whiten, A., eds. 1988. *Machiavellian intelligence: Social expertise and the evolution of intellect in monkeys, apes, and humans*. Oxford, UK: Clarendon Press.

Calvin, W. H. 1996. *The cerebral code*. Cambridge, MA: MIT Press.

Campbell, D. T. 1960. Blind variation and selective retention in creative thought as in other knowledge processes. *Psychological Review* 67: 380–400.

——. 1990. Epistemological roles for selection theory. In *Evolution, cognition and realism*, ed. N. Rescher, pp. 1–19. Lanham, MD: University Press of America.

Casey, E. S. 1987. *Remembering: A phenomenological study*. Bloomington and Indianapolis: Indiana University Press.

Cermak, L. S. 1987. Models of memory loss in Korsakoff and alcoholic patients. In *Neuropsychology of alcoholism*, ed. O. A. Parsons, N. Butters, and P. E. Nathan, pp. 207–226. New York: Guilford Press, Inc.

Changeux, J. P. 1985. *Neuronal man: The biology of the mind.* New York: Pantheon.

Changeux, J. P., and Dehaene, S. 1989. Neuronal models of cognitive functions. *Cognition* 33: 63–109.

Child, I. L. 1985. Psychology and anomalous observations. The question of ESP in dreams. *American Psychologist* 40 (11): 1219–30.

Chomsky, N. 1988. *Language and problems of knowledge: The Managua lectures.* Cambridge, MA: MIT Press.

Coomaraswamy, A. K. 1977. Recollection, Indian and Platonic. In *Coomaraswamy* 2: *Selected papers, metaphysics*, ed. R. Lipsey. Princeton, NJ: Bollingen Series, Princeton University Press, pp. 49–65.

Crick, F., and Mitchison, G. 1986. REM sleep and neural nets. *The Journal of Mind and Behavior* 7 (2 and 3): 229 –250.

Critchley, M. 1979. *The divine banquet of the brain and other essays.* New York: Raven Press.

Cummings, J. L. 1993. Frontal-subcortical circuits and human behavior. *Archives of Neurology* 50: 873–80.

Cziko, G. 1995. *Without miracles: Universal selection theory and the second Darwinian revolution.* Cambridge, MA: MIT Press.

Dagenbach, D., and Carr, T. H. 1994. Inhibitory processes in perceptual recognition: Evidence for a center-surround attentional mechanism. In *Inhibitory processes in attention, memory and language*, ed. D. Dagenbach and T. H. Carr, pp. 327–359. San Diego: Academic Press.

Darden, L., and Cain, J. A. 1989. Selection type theories. *Philosophy of Science* 56: 106–129.

Darwin, C. 1859. *On the origins of species by means of natural selection or the preservation of favored races in the struggle for life.* London: John Murray.

Dawkins, R. 1976. *The selfish gene.* Oxford, UK: Oxford University Press.

Deleuze, G. 1991. *Bergsonism.* New York: Zone Books.

Dempster, F. N., and Brainerd, C. J., eds. 1995. *Interference and inhibition in cognition.* San Diego: Academic Press.

Dennett, D. 1995. *Darwin's dangerous idea: Evolution and the meanings of life.* New York: Touchstone.

Dessauer, F. 1983. Technology in its proper sphere. In *Philosophy and Technology. Readings in the philosophical problems of technology*, ed. C. Mitcham and R. Mackey, pp. 317–334. New York: The Free Press.

Edelman, G. M. 1987. *Neural Darwinism.* New York: Basic Books.

Edman, I. 1928. *The works of Plato: The Jowett translation.* New York: Simon and Schuster.

Eliade, M. 1963. *Myth and Reality.* New York: Harper and Row.

Eysenck, H. J., 1994. Personality: Biological foundations. In *The neuropsychology of individual differences,* ed. P. A. Vernon, pp. 151–207. New York: Academic Press.

Fodor, J. A. 1983. *The modularity of Mind: An essay on faculty psychology.* Cambridge, MA: MIT Press.

Foulkes, D. 1985. *Dreaming: A cognitive-psychological analysis.* Hillsdale, NJ: Lawrence Erlbaum Associates.

Fuster, J. M. 1989. *The prefrontal cortex: Anatomy, physiology and neuropsychology of the frontal lobe.* 2nd ed. New York: Raven Press.

Gazzaniga, M. S. 1992. *Nature's mind: the biological roots of thinking, emotions, sexuality, language and intelligence.* New York: Basic Books.

Gernsbacher, M. A., and Faust, M. 1995. Skilled suppression. In *Interference and inhibition in cognition,* ed. F. N. Dempster and C. J. Brainerd, pp. 295–327. San Diego: Academic Press.

Gillin, J. C.; Buchsbaum, M. S.; and Wu, J. C. 1993. Cerebral metabolism. In *Encyclopedia of Sleep and Dreaming,* M. A. Carskadon editor-in-chief, pp. 96–98. New York: Macmillan Publishing.

Goldberg, G. 1987. From intent to action: Evolution and function of the premotor systems of the frontal lobe. In *The frontal lobes revisited,* ed. E. Perceman, pp. 273–306. New York: IRBN Press.

Goldman-Rakic, P. S. 1994. The issue of memory in the study of prefrontal function. In *Motor and cognitve functions of the prefrontal cortex,* ed. A-M Thierry et al., pp. 112–121. Berlin: Springer-Verlag.

Gray, J. A. 1991. The neuropsychology of temperament. In *Explorations in temperament: International perspective on theory and measurement,* ed. J. Strelau and A. Angleitner, pp. 115–130. New York: Plenum Press.

Greenwald, A. G. 1981. Self and memory. *Psychology of Learning and Motivation* 15: 202–233.

Grossberg, S. 1980. How does the brain build a cognitive code? *Psychological Review* 87: 1–51.

Hale, N. G. 1971. *James Jackson Putnam and psychoanalysis.* Cambridge, MA: Harvard University Press.

Hall, C. S., and Van de Castle, R. L. 1966. *The content analysis of dreams.* New York: Appleton-Century-Crofts.

Hartmann, E. 1991. *Boundaries in the mind: A new psychology of personality.* New York: Basic Books.

——. 1996. Outline for a theory on the nature and functions of dreaming. *Dreaming* 6: 147–170.

Heylighen, F. 1995. *Memetic selection criteria.* http//pespmc1. vub. ac. be/memselc. html.

Hobson, J. A. 1988. *The dreaming brain.* New York: Basic Books.

Houghton, G., and Tipper, S. P. 1994. A model of inhibitory mechanisms in selective attention. In *Inhibitory processes in attention, memory and language,* ed. D. Dagenbach and T. H. Carr, pp. 53–112. San Diego: Academic Press.

Hull, D. 1980. Individuality and selection. *Annual Review of Ecology and Systematics,* 11: 311–332.

Jacoby, L.; Kelley, C. M.; Brown, J.; and Jaechko, J. 1989. Memory attributions. In *Varieties of memory and consciousness: Essays in honor of Endel Tulving,* ed. H. L. Roediger and F. I. M. Craik, pp. 391–422. Hillsdale, NJ: Lawrence Erlbaum.

James, W. 1890/1981. *The principles of psychology.* Cambridge, MA: Harvard University Press.

——. 1909. *A pluralistic universe.* New York: Longman, Green and Co.

——. 1956. *The will to believe and other essays in popular philosophy.* New York: Dover Publications, Inc.

Jerne, N. K. 1985. The generative grammar of the immune system. Nobel lecture. *Science* 229: 1057–1059.

Johnson, M. K.; Foley, M.; Suengas, A. G.; and Raye, C. L. 1988. Phenomenal characteristics of memories for perceived and imagined autobiographical events. *Journal of Experimental Psychology: General* 4: 371–376.

Kahnemann, D. 1973. *Attention and effort.* Englewood Cliffs, NJ: Prentice Hall.

Kant, Immanuel. 1929. *Critique of pure reason.* N. Kemp Smith, trans. New York: St. Martin's Press.

——. 1956. *Critique of practical reason.* L. W. Beck, trans. Indianapolis: Bobbs-Merrill.

Kihlstrom, J. F.; Cantor, N.; Albright, J. S.; Chew, B. R.; Klein, S. B.; and Niedenthal, P. M. 1988. Information processing and the study of the self. *Advances in Experimental Social Psychology* 21: 145–178.

Klinger, E. 1978. Modes of normal conscious flow. In *The stream of consciousness,* ed. K. S. Pope and J. L. Singer. New York: Plenum.

Krell, D. F. 1990. *Of memory, reminiscence and writing: On the verge.* Bloomington and Indianapolis: Indiana University Press.

LaBerge, D. 1995. *Attentional processing.* Cambridge, MA: Harvard University Press.

Lewontin, R. C. 1982. Organism and environment. In *Learning, development and culture,* ed. H. C. Plotkin, pp. 151–170. Chicester: John Wiley and Sons.

Lhermitte, F. 1986. Human autonomy and the frontal lobes. II. Patient behavior in complex and social situations. The "environmental dependency syndrome." *Annals of Neurology* 19: 335–43.

Lieberman, P. 1984. *The biology and evolution of language.* Cambridge, MA: Harvard University Press.

Loftus, E. F.; Feldman, J.; and Dashiell, R. 1995. The reality of illusory memories. In *Memory distortion: How minds, brains, and societies reconstruct the past,* ed. D. L. Schacter, J. T. Coyle, G. D. Fischbach, M. M. Mesulam and L. E. Sullivan, pp. 47–68. Cambridge, MA: Harvard University Press.

Magnussen, S., and Dyrnes, S. 1994. High-fidelity perceptual long term memory. *Psychological Science* 52: 99–101.

Malcolm, N. 1977. *Memory and mind.* Oxford, UK: Oxford University Press.

Marks, C. E. 1986. *Comissurotomy, consciousness and unity of mind.* Cambridge, MA: MIT Press.

McCarley, R. W. 1989. The biology of dreaming sleep. In *Principles and practice of sleep medicine,* ed. M. H. Kryger, T. Roth, and W. C. Dement, pp. 173–183. Philadelphia: Saunders.

McGlynn, S. M., and Schacter, D. L. 1989. Unawareness of deficits in neuropsychological syndromes. *Journal of Clinical and Experimental Neuropsychology* 11: 143–205.

Mehler, J., and Fox, R., eds. 1985. *Neonate cognition: Beyond the blooming buzzing confusion.* Hillsdale, NJ: Lawrence Erlbaum Associates.

Moffitt, A.; Kramer, M.; and Hoffman, R., eds. 1993. *The functions of dreaming.* Albany: State University of New York Press.

Mumford, L. 1966. *The myth of the machine. Technics and human development.* New York: Harcourt Brace and World, Inc.

Myers, G. E. 1986. *William James. His life and thought.* New Haven, CT: Yale University Press.

Näätänen, R. 1982. Processing negativity: An evoked potential reflection of selective attention. *Psychological Bulletin* 92: 605–640.

Neimeyer, G. J., and Metzler, A. E. 1994. Personal identity and autobiographical recall. In *The remembering self. Construction and accuracy in the self narrative*, ed. U. Neisser and R. Fivush, pp. 105–135. Cambridge, U.K.: Cambridge University Press.

Nigro, G., and Neisser, U. 1983. Point of view in personal memories. *Cognitive Psychology* 15: 467–482.

Norman, D. A. 1968. Toward a theory of memory and attention. *Psychological Review* 75: 522–536.

Nyberg, L.; Cabeza, R.; and Tulving, E. 1996. PET studies of encoding and retrieval: The HERA model. *Psychonomic Bulletin and Review* 2: 134–147.

O'Leary, D. D. M. 1992. Development of connectional diversity and specificity in the mammalian brain by the pruning of collateral projections. *Current Opinion in Neurobiology* 2: 70–77.

Oscar-Berman, M.; McNamara, P.; and Freedman, M. 1991. Delayed response tasks: Parallels between experimental ablation studies and findings in patients with frontal lesions. In *Frontal lobe function and injury*, ed. H. S. Levin and H. M. Eisenberg, pp. 230–255. New York: Academic Press.

Pandya, D. N., and Barnes, C. L. 1987. Architecture and connections of the frontal lobe. In *The frontal lobes revisited*, ed. E. Perecman, pp. 41–72. New York: IRBN Press.

Piatelli-Palmarini, M. 1989. Evolution, selection and cognition: From "learning" to parameter setting in biology and in the study of language. *Cognition* 31 (1): 1–44.

Pinker, S. 1994. *The language instinct*. New York: Morrow.

Plotkin, H. 1994. *Darwin machines and the nature of knowledge*. Cambridge, MA: Harvard University Press.

Pope, K. S., and Singer, J. L. 1977. Regulation of the stream of consciousness: Towards a theory of ongoing thought. In *Consciousness and self-regulation*, vol. 2, ed. G. Schwartz and D. Shapiro, pp. 101–137. New York: Holt, Rinehart and Winston.

Popper, K. 1963. *Conjectures and refutations*. London: Routledge and Kegan Paul.

Pribram, K. H., and McGuinness, D. 1975. Arousal, activation and effort in the control of attention. *Psychological Review* 82: 116–149.

Quattrone, G., and Tversky, A. 1984. Causal versus diagnostic contingencies: On self-deception and the voter's illusion. *Journal of Personality and Social Psychology* 46: 237–248.

Raaijmakers, J. G. W., and Shiffrin, R. M. 1981. Search of associative memory. *Psychological Review* 88 (2): 93–135.

Rakic, P.; Bourgeois, J-P.; Eckenhoff, M. F.; Zecevic, N.; and Goldman-Rakic, P. S. 1986. Concurrent overproduction of synapses in diverse regions of the primate cerebral cortex. *Science* 232: 232–235.

Reeke, G. N., and Sporns, O. 1990. Selectionist models of perceptual and motor systems and implications for functionalist theories of brain function. In *Emergent computation: Self organizing, collective, and cooperative phenomena in natural and artificial computing networks*, ed. S. Forrest, pp. 347–364. Cambridge, MA: MIT Press.

Richardson-Klavehn, A., and Bjork, R. A. 1988. Measures of memory. *Annual Review of Psychology*, 39: 475–543.

Rindos, D. 1985. Darwinian selection, symbolic variation, and the evolution of culture. *Current Anthropology* 26 (1): 65–88.

Ross, M., and Buehler, R. 1994. Creative remembering. In *The remembering self: Construction and accuracy in the self narrative*, ed. U. Neisser and R. Fivush, pp. 205–235. Cambridge, UK: Cambridge University Press.

Sacks, O. 1992. A neurologist's notebook. A surgeon's life. *The New Yorker*, March 16, pp. 85–94.

Sanides, F. 1969. Comparative architectonics of the neocortex of mammals and their evolutionary significance. *Annals of the New York Academy of Sciences* 167: 404–423.

Schacter, D. L. 1996. *Searching for memory. The brain, the mind and the past.* New York: Basic Books.

Schmid, M., and Wuketits, F. M., eds. 1987. *Evolutionary theory in social science*. Dordrecht: D. Reidel Publishing Co.

Semon, R. 1909/1923. *Mnemic psychology*. London: George Allen and Unwin.

Singer, J. L. 1984. The private personality. *Personality and Social Psychology Bulletin* 10 (1): 7–30.

——. 1993. Experimental studies of ongoing conscious experience. In *Experimental and theoretical studies of consciousness*, ed. G. R. Bock and J. Marsh, pp. 100–122. West Sussex, UK: John Wiley and Sons.

Singer, J. L., and Bonanno, G. A. 1990. Personality and private experience: Individual variations in consciousness and in attention to subjective phenomena. In *Handbook of Personality: Theory and Research*, ed. L. A. Pervin, pp. 419–444. New York: Guilford Press.

Sober, E. 1984. *The nature of selection.* Cambridge, MA: MIT Press.

Sporns, O. 1994. Selectionist and instructivist ideas in neuroscience. In *Selectionism and the brain*, ed. O. Sporns and G. Tononi, *International Review of Neurobiology* 37: 4–26. New York: Academic Press.

Sporns O., and Tononi, G., eds. 1994. Selectionism and the brain. *International Review of Neurobiology* 37. New York: Academic Press.

Squire, L. R. 1994. Memory and forgetting: Long term and gradual changes in memory storage. In *Selectionism and the brain*, ed. O. Sporns and G. Tononi, *International Review of Neurobiology* 37: 243–270. New York: Academic Press.

Steriade M. 1989. Brain electrical activity and sensory processing during waking and sleep states. In *Principles and practice of sleep medicine*, ed. M. H. Kryger, T. Roth, and W. C. Dement, pp. 86–103. Philadelphia: Saunders.

Stich, S. 1991. *The fragmentation of reason.* Cambridge, MA: MIT Press.

Swinney, D. A., and Prather, P. 1989. On the comprehension of lexical ambiguity by young children: Investigations into the development of mental modularity. In *Resolving semantic ambiguity*, ed. D. S. Gorfein, pp. 225–238. New York: Springer-Verlag.

Symons, D. 1993. The stuff that dreams aren't made of: Why wake-state and dream-state sensory experiences differ. *Cognition* 47: 181–217.

Tipper, S. P. 1992. Selection for action: The role of inhibitory mechanisms. *Current Directions in Psychological Science* 1: 105–109.

Treisman, A. M. 1969. Strategies and models of selective attention. *Psychological Review* 95: 15–48.

Trivers, R. 1985. *Social evolution.* Menlo Park, CA: Benjamin Cummings.

Tucker, D. M., and Williamson, P. A. 1984. Asymmetric neural control systems in human self-regulation. *Psychological Review* 91 (2):185–215.

Tulving, E. 1983. *Elements of episodic memory.* Oxford, UK: Oxford University Press.

Van de Castle, R. L. 1994. *Our dreaming mind.* New York: Ballantine Books.

Vico, Giambattista 1744/1975. *The New Science of Giambattista Vico.* Ed. Thomas Goddard Bergin and Max Harold Fisch. Ithaca: Cornell University Press, 1975.

Ward, P. J. 1989. A model of end-directedness. In *Issues in evolutionary epistemology*, ed. K. Hahlweg and C. A. Hooker, pp. 357–390. Albany: State University of New York Press.

Wheeler, M. A.; Stuss, D. T.; and Tulving, E. 1997. Toward a theory of episodic memory: The frontal lobes and autonoietic consciousness. *Psychological Bulletin* 121 (3): 331–354.

Winson, J. 1985. *Brain and psyche: The biology of the unconscious*. New York: Doubleday.

Yates, F. A. 1966. *The art of memory*. Chicago: University of Chicago Press.

Young, J. Z. 1979. Learning as a process of selection and amplification. *Journal of the Royal Society of Medicine* 72: 801–814.

Yourcenar, M. 1954. *Memoirs of Hadrian*. New York: Farrar, Straus and Giroux.

Zuckerman, M. 1994. *Behavioral expressions and biosocial bases of sensation seeking*. Cambridge, UK: Cambridge University Press.

Index

About the Author

PATRICK McNAMARA is Assistant Professor of Behavioral Neurosciences at Boston University School of Medicine, and Research Director at the Partnership for Organ Donation in Boston. He has a background in behavioral neuroscience, and his research interests include memory, language, and frontal lobe disorders. He has published several scholarly articles and book chapters.

Edwards Brothers Malloy
Thorofare, NJ USA
July 2, 2013